# Disaster Management

## A Disaster Manager's Handbook

Food and Agriculture
Organization of the United Nations

Reprinted in India by:

**UNITED BOOK PRINTS**
(Imprint of Scientific Publishers, INDIA)
4806/24, Ansari Road, Daryaganj,
New Delhi 110 002, INDIA
E-mail: info@scientificpub.com
Website : www.scientificpub.com

Print 2026



ISBN : 978-93-90749-99-7



Printed in India

# Contents

## PART III – MAJOR FACTORS PRIOR TO DISASTER IMPACT



## PART IV – RESPONSE TO DISASTER IMPACT

## PART V – MAJOR POST-IMPACT FACTORS



## PART VI – DISASTER MANAGEMENT SUPPORT REQUIREMENTS

## APPENDIXES

# LIST OF ACRONYMS

| | |
|---|---|
| **ADPC** | Asian Disaster Preparedness Center |
| **AIDAB** | Australian International Development Assistance Bureau |
| **AODRO** | Australian Overseas Disaster Response Organization |
| **ASEAN** | Association of Southeast Asian Nations |
| **CCG** | Central Control Group |
| **DMC** | developing member country |
| **EOC** | emergency operations center |
| **GIS** | Geographic Information System |
| **NATO** | North Atlantic Treaty Organisation |
| **NDC** | National Disaster Council |
| **NDMO** | National Disaster Management Office |
| **NEOC** | National Emergency Operations Center |
| **NGO** | nongovernment organization |
| **OGC** | Operations Control Group |
| **PIDP** | Pacific Islands Development Program |
| **UK** | United Kingdom |
| **UNDRO** | United Nations Disaster Relief Office |
| **UNHCR** | United Nations High Commission for Refugees |
| **US** | United States |
| **USAID** | United States Agency for International Development |

# Preface

This book stems from a regional technical assistance project that was approved by the Asian Development Bank in January 1990 under the title "Regional Study of Disaster Mitigation." The objectives of the study were (i) to enhance the awareness of the ADB's developing member countries (DMCs) of the need for an active developmental effort to mitigate the impact of natural disasters; (ii) to produce authoritative documentation on disaster management practices in selected DMCs; (iii) to improve the understanding of the techniques of disaster mitigation through structural and nonstructural mitigation practices; and (iv) to provide a forum for an exchange of experiences in and views on designing and implementing national strategies for mitigating the impact of natural disasters on economic development, and the formulation of national and regional strategies.

Financed by a technical assistance grant, the study included a regional seminar held in October 1990 for representatives of DMCs concerned and dealing with the formulation of country-specific disaster management responses. This seminar led to two publications. The first was a volume entitled *Disaster Mitigation in Asia and the Pacific* which was published in July 1991, drawing on the lessons of experience and current disaster management practices in Bangladesh, Nepal, Philippines, and countries in the South Pacific, with a view to designing a general response mechanism for addressing identified gaps in existing disaster mitigation practices in the region.

The present volume is the second product of the exercise. It is intended to serve as a ready reference guide for those actively involved in DMCs with management of natural disasters before, during, and after the disaster situations. It is a further example of the ADB's commitment to support efforts in disaster mitigation in the region. To date, no handbook has been published specifically for the guidance of agencies and officials of disaster-prone developing countries. This handbook is intended to fill that gap, and as such, complements the first volume.

This handbook reflects the ADB's ongoing concern for sustaining development process in the face of the periodic disasters that occur in Asia and the Pacific region. It is hoped that this volume will serve as a practical guide for disaster managers and enable them to assume an effective role to benefit millions of people who, without efficient disaster management, would be left defenseless against the vagaries of nature.

**A.I. AMINUL ISLAM**
Chief
Development Policy Office

Asian Development Bank
February 1992

# Foreword

Most of the world's natural disasters occur in Asia and the Pacific. They cause enormous destruction and human suffering in the Asian Development Bank's (ADB) developing member countries (DMCs).

Environmental degradation, which is often a result of economic development and associated human settlement patterns that ignore appropriate resource management, can increase a country's vulnerability to natural hazards and exacerbate the impacts.

As a development institution, ADB is quite concerned about the rising incidence and impact of natural disasters in Asia and the Pacific region. Losses from natural disasters reduce the pace of sustained economic development and often lead to a heavy drain on available resources, diverting them from pursuing developmental aims. Preparing for disaster situations and, if possible, preventing them is considered by many as an integral part of development planning focusing on achieving sustainable growth, poverty reduction, and environmental balance. ADB shares this view.

In addition to utilizing its normal project lending operations for incorporating disaster-mitigation measures into developmental activities and financing disaster-rehabilitation projects, ADB may rely on its technical assistance facility for promoting disaster mitigation. Such assistance is available to its DMCs for the formulation of disaster-mitigation plans, devising mitigation strategies, and designing measures to carry them out. This type of ADB assistance is available on either a regional, country, or project level.

An example of technical assistance on the regional level is the present effort by ADB to assist its DMCs to gain a better understanding of disaster risks and vulnerability to natural hazards, and to focus their attention on disaster management issues. This Handbook, as well as the first volume *Disaster Mitigation in Asia and the Pacific* published in July 1991, are products of the ADB-financed Regional Seminar on Disaster Mitigation, October 1990.

The principal concept underlying this handbook is that modern disaster management is very much an ongoing national requirement, important to governments and people alike. It has special significance today because of increasing dangers to the world environment posed by natural hazards and their effects on economic development. To be effective, disaster management should be implemented as a comprehensive and continuous activity, not as a periodic reaction to individual disaster circumstances. Consequently, national officials who are charged with disaster management responsibilities have to deal with a wide range of policy, planning, organizational, operational, and other matters. It is clearly beneficial for them to have access to relevant guidelines and advice, especially in the form of a ready-reference handbook.

The handbook draws upon disaster management practices in the region and endeavors to relate realistically to the needs of national disaster managers. It covers the broad disaster management field and offers guidelines on the major segments that constitute that field. The handbook does not attempt to deal with highly specialized aspects of disasters such as detailed economic, technical, and sociological issues. These are more the concern of specialists who are normally available to provide the relevant advice. The reader should find the handbook simple and straightforward to use. Individual chapters are as self-contained as possible, with adequate cross-references where necessary. Its content is not dogmatic; instead it takes an advisory line, including options and case material references to help disaster managers in resolving their local needs. This contributes to the application of past experiences accrued within the region and, where appropriate, elsewhere.

ADB's chief consultant and lead author for this handbook was Air Vice Marshal Nick Carter. In putting these guidelines together, he has drawn upon more than 50 years of experience: first, during the Second World War when civil defense was, for many people, essential to survival; second, in civil unrest which is a problem that continues to plague the region; third, in strategic planning for the ultimate disaster, nuclear war; fourth, as director for 10 years of the Australian Counter Disaster College; fifth, for another 12 years, as an international disaster consultant undertaking numerous missions throughout Asia and the Pacific, and sixth, but by no means least, as a victim of one of the worst bushfires ever to strike Australia.

Thanks are due to the Director and staff of the Asian Disaster Preparedness Center at the Asian Institute of Technology, Bangkok, Thailand, our consultants for the overall exercise, who also saw to the timely completion of the handbook; to the Governments of the Cook Islands, Papua New Guinea, and Queensland that have generously permitted the reproduction of their disaster legislation here; and the panel of experts who reviewed an earlier draft of the handbook: The Honorable Secretary Dr. Mita Pardo de Tavera (Philippines), Leith Anderson (Papua New Guinea), Hugh Brammer (United Kingdom), Nelson Delailomaloma (Fiji Islands), Charitha Ratwatte (Sri Lanka), and George Ritchie (United Kingdom). On the ADB's side, the work was initiated and coordinated by Dr. Werner M. Schelzig of the Development Policy Office. Their contributions are gratefully acknowledged.

The publication of this handbook brings ADB activities under TA 5353-REG, Regional Study of Disaster Mitigation, to an end. I am pleased to note that judging from the overwhelmingly positive response we have had to the regional seminar and the first volume published, we have reason to believe that the findings of TA 5353-REG, as expressed in the two publications at hand, form a solid base for further action. ADB stands ready to continue its dialogue on disaster mitigation with its DMCs and in close cooperation with the international community, with a view to making disaster management an integral part of development planning in the region and a regular practice.

**WILLIAM R. THOMSON**
Vice President (Operations)

Asian Development Bank
February 1992

# Introductory Notes

## Concept of the Handbook

The concept of this handbook is that modern disaster management is very much an ongoing national requirement which is important to governments and peoples alike. It has special significance today because of increasing dangers to the world environment.

To be effective, therefore, disaster management needs to be implemented as a comprehensive and continuous activity, not as a periodic reaction to individual disaster circumstances. Consequently, national representatives who are charged with disaster management responsibilities have to deal with a wide range of policy, planning, organizational, operational, and other matters. It is therefore clearly beneficial if they can have access to relevant guidelines and advice, especially in the form of a ready-reference handbook.

## Aim

The aim of the handbook is to provide a reference and guide to those who, within governments and outside, may hold key responsibilities for dealing with disaster.

Thus, its underlying purpose is to assist disaster managers in attaining standards of professionalism that will enable them to deal with day-to-day disaster management circumstances and with actual disaster situations.

## Focus

The focus of the handbook is primarily on countries of Southeast Asia and the Pacific region. It draws on factual disaster management activities within this region and endeavors to relate realistically to the needs of disaster managers within this region. Thus, the handbook is a practical reference, not a collection of theoretical propositions.

Despite this regional focus, however, the handbook follows disaster principles and practices that are clearly applicable to other areas and circumstances.

## Scope and Limitations

As a practical reference, the handbook is designed to cover the broad disaster management field and to offer guidelines on the major segments which constitute that field.

The handbook does not attempt to deal with highly specialized aspects that are associated with disasters such as detailed economic, technical, and sociological issues. These are primarily the concern of existing specialists, who are normally available to provide relevant advice.

## Format and Style

The handbook is presented in a practical, ready-reference style. It is divided into major parts and its chapters aim to be as self-contained as possible, with adequate cross-references where necessary. Thus, the reader should find the handbook simple and straightforward to use.

Also, the handbook endeavors to avoid a dogmatic approach. Instead, it takes an advisory line, including the offer of options and case material references to assist disaster managers in resolving their individual local needs. This, it is believed, helps in applying and utilizing past experience.

Other considerations of format and style are:

- The handbook is intended to help those for whom English may be a second language.
- The format used should enable the handbook to be easily translated into other languages, where this is necessary.
- Definitions used in the handbook have been selected primarily for their clarity of description.
- The guidelines provided in the handbook are designed so that they may be easily adapted to various local circumstances.

## The Definition of Disaster

There are many different definitions of disaster. Most such definitions tend to reflect the following characteristics:

- Disruption to normal patterns of life. Such disruption is usually severe and may also be sudden, unexpected, and widespread.
- Human effects such as loss of life, injury, hardship, and adverse effect on health.
- Effects on social structure such as destruction of or damage to government systems, buildings, communications, and essential services.
- Community needs such as shelter, food, clothing, medical assistance, and social care.

Two dictionary definitions are:

- *Concise Oxford Dictionary*
  Sudden or great misfortune, calamity.
- *Webster's Dictionary*
  A sudden calamitous event producing great material damage, loss, and distress.

The variety of definitions in use and the need for individual organizations to frame their own definitions, has been borne in mind during the compilation of this handbook. The definition used in the handbook is:

> An event, natural or man-made, sudden or progressive, which impacts with such severity that the affected community has to respond by taking exceptional measures.

In relation to the definition of disaster, it has also been taken into account that disaster management is essentially a dynamic process. It encompasses the classical management functions of planning, organizing, staffing, leading, and controlling. It also involves many organizations which must work together to prevent, mitigate, prepare for, respond to, and recover from the effects of disaster. Disaster management is defined as:

> An applied science which seeks, by the systematic observation and analysis of disasters, to improve measures relating to prevention, mitigation, preparedness, emergency response and recovery.

## Practical Application of Disaster Definition

There is an important practical application of the definition of disaster in disaster management. Such definition helps provide a common concept and theme throughout disaster management activities. Thus, the chosen definition is valuable for purposes of policy, organization, planning, and legislation.

It is suggested that individual nations and organizations should choose a definition that is most suitable for their purposes and apply it accordingly.

## The Disaster Management Cycle

A standard cycle of disaster events has been used throughout the handbook. It is illustrated below:

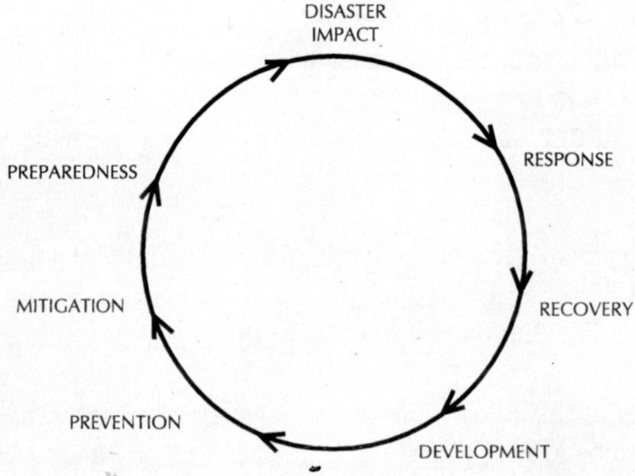


This cycle can be represented in more detail, as explained in Chapter 5.

The cycle can be, and often is portrayed in other forms. Moreover, alternative terminology may be used. The important factor is that the format should indicate that disaster and its management is a continuum of interlinked activities; it is not a series of events which start and stop with each disaster occurence.

## Government Structures

It is recognized that the structure of government may differ as between countries. For the purpose of this handbook, therefore, the following terms are used to differentiate generally between the various levels:

| *Term* | *Levels of Government* |
|---|---|
| National Level | National or central |
| Intermediate Level | Provincial, state, or regional |
| Community level | Municipality, local government, or village |

## Relevance to Other Handbooks

During compilation of this handbook, it has been borne in mind that other disaster-related handbooks exist. Most of them directly relate to the specific roles and requirements of their issuing organizations such as:

- Red Cross
- United Nations High Commission for Refugees
- United Nations Children's Fund, and
- United Nations Development Programme/United Nations Disaster Relief Office

As far as practicable, and certainly where major considerations are concerned, this handbook has endeavored to maintain reasonable compatibility with these co-references.

# PART I

# Basic Disaster Management Aspects

*Chapter 1*

# The Significance of Disaster

## Purpose

1. The purpose of this chapter is to remind the reader of the significance of disaster in today's environment. This is in order that individual disaster management authorities and officials may assess this significance in relation to their own circumstances.

## Introduction

2. The significance of disaster in today's environment sometimes comes under question. Why do we need to bother so much? After all, disaster has been with us as long as recorded history, and presumably even longer. Generations of people have had to withstand disaster. They have suffered the consequences and recovered from them, and life has continued. Basically, this is true. However, certain factors need to be considered in relation to modern challenges which face disaster management.

## The Traditional Disaster Threat

3. There has not been very much reduction in what might be called the traditional disaster threat. Most of the old problems remain, as threatening as ever. Natural phenomena such as earthquakes, cyclones, volcanic eruptions, tsunamis, wildfires, floods, landslides, and drought still persist. So do their basic man-made counterparts such as major accidents. These disasters continue to cause grievous human casualties, economic and social loss, and damage to the environment. It is certainly true that we have learned to cope with these problems to some extent. But we have neither eliminated nor contained them. So while we may have modified their effects in various ways, they continue to inflict unacceptable pressure on a world population which, in terms of total subsistence, is already finding it difficult to make ends meet.

4. In fact, some long-standing threats have grown more severe. For example, the risk from air disaster was insignificant in the 1920s. Few aircraft were in the air and a collision between two of them would have killed only a

handful of people at most. Now, as the 20th century comes to a close, the air disaster risk has increased enormously. More and more aircraft fill the already overcrowded airspaces of the world, especially around capital cities. A collision between two of them can be catastrophic. In 1977, a collision in the Azores between two passenger-carrying jets resulted in the deaths of 561 people. One aircraft was still on the ground at the time. In 1988, 270 people lost their lives following the terrorist sabotage of an airliner over Scotland.

5. With some of the other long-standing threats, we ourselves have added to the risks. An increasing population alone has forced people to live in disaster-prone areas which, previously, would not have been regarded as habitable. This fact tends to apply particularly to developing countries. For instance, human settlement has been allowed to develop not only in flood-prone areas of major river systems, but also on low atoll islands which are subject to inundation from the sea. The chain effects are sometimes disturbing. Sea inundation can cause over-salinity of crop-growing land, food shortages follow, and populations may be forced into crises of subsistence or even famine—leading perhaps to migration and refugee problems.

6. What is often seen as progress can, in fact, represent a backward step. In some cyclone-prone countries, traditional building construction—designed to cope with high winds—has been modified. Increased population has led to the need for more water, so tin roofing has been introduced to enhance water collection. But a piece of iron roofing, ripped off a house by cyclonic winds, and moving at 100 kilometers an hour, is a very lethal weapon that is just as capable of killing people as any modern weapons system. Traditional building materials posed no such threat because most were made of lightweight wood or similar substances.

## The New Disaster Threats

7. A second factor which bears upon today's situation is that new disaster threats have developed, particularly since World War II.

8. Increased social violence has drastically affected many nations and communities. Highjacking, terrorism, civil unrest, and conflict with conventional arms have become commonplace. These have sometimes inflicted intolerable burdens on government and societies whose existence is already precarious because of poor economic and social conditions. This, in turn, has produced additional strains on international assistance sources, thus diluting global counter-disaster efforts and capabilities.

9. New threats have also come from what are generally termed hazardous materials or substances. The 1985 tragedy of Bhopal in India ranks paramount in this category with an estimated death toll of 2,500 and 100,000 affected in various ways. But the Bhopals of this world are, in many ways, only the highly publicized tip of this particular disaster iceberg. Hazardous materials are shifted around the transport systems of the world in increasing amounts. They are sometimes dumped in areas which are vital to the world's future. These materials can constitute a disaster threat that is potentially comparable to those posed by many natural phenomena.

10. The threat from atomic and nuclear sources poses yet another modern problem for disaster management. The 1986 Chernobyl nuclear power plant explosion in the former Soviet Union highlighted the extent and severity of this problem. Apart from those killed and affected by radiation sickness, some 135,000 people had to be evacuated from the area. Radioactive effects from the disaster were measured as far away as 1,600 miles and more. These peacetime nuclear problems are likely to persist in a world that increasingly searches for new and greater sources of energy.

11. While the threat from nuclear accidents is disturbing enough, the disaster management problems arising from a possible nuclear war are almost beyond comprehension. The possibility of a global nuclear war may have receded in recent years, but the possibility of nuclear weapons being used in some lesser form of conflict cannot be disregarded altogether. Also, it would be unwise to rule out entirely the use of such weapons by fanatical extremist factions. Furthermore, if a country was not directly involved in such nuclear conflict or terrorism, it could well suffer severely from radioactive side effects.

12. In sum, therefore, new disaster threats can contain some unwelcome characteristics in that they may have extremely far-ranging effects and, at the same time, be difficult to counter.

## The Geography of Disaster

13. A third factor concerns what might be called the geography of disaster. It has often been pointed out that most of the world's worst disasters tend to occur between the Tropic of Cancer and the Tropic of Capricorn and that, coincidentally, this area contains the poorer countries.[1] Of course, a major significance of this is that such countries find themselves facing repeated setbacks to progress. Indeed, some countries seem destined to remain within

---

[1] See, for instance, Cuny, Frederick C. 1983. *Disasters and Development*. New York: Oxford University Press.

the developing category primarily because of the severity and magnitude of their disasters. Seen in this light, therefore, disaster can be a strong aggravating factor in the differences between wealthy and poor nations—that is, between the "haves" and the "have-nots."

## The Modern Loss Factor

14. The fourth factor is the relationship between contemporary disaster threat and the losses it may impose. The simple fact is that the more nations develop and the more assets they build up, the more they stand to lose. It follows that any action which can be taken to reduce disaster-related loss must be seen as logical and desirable in cost–benefit terms. This applies to all countries—rich or poor—and it underlines the need for all countries to try to develop and maintain an effective disaster management capability that is appropriate to their needs. It also underlines the necessity of coordinated international action to strengthen all aspects of disaster management, wherever this is possible. In this regard, the initiation under the auspices of the United Nations of an International Decade for Natural Disaster Reduction during the 1990s must be seen as a progressive and highly commendable step.

## Major Aspects of Significance

15. Following from the foregoing paragraphs in this chapter, the significance of disaster can perhaps best be summarized in global, national, and practical disaster management terms.

16. In global terms, unless disaster can be mitigated and managed to the optimum extent possible, it will continue to have a dominating effect on the future. The world is already facing a range of environmental and subsistence crises. Disaster mitigation should be regarded as an important tool in successfully coping with these crises. Moreover, the political, economic, and social stability of the world depends significantly on bridging the gap between developing and developed nations. Mitigating and containing the effects of disaster on the developing nations now and in the future is an important asset toward bridging this gap. The continuation and enhancement of international disaster assistance is also a key factor. Correctly applied, such assistance can help provide a desirable bonding between nations, and thus, produce welcome and beneficial long-term results.

17. In national terms, the impact of disaster usually results in two major setbacks:

- First, the direct loss of existing national assets in various forms.

- Second, the diversion of national resources and effort away from ongoing subsistence and development to achieve satisfactory recovery.

This indicates that nations need to develop a comprehensive approach to disaster management. Only by such an approach can they hope to deal effectively with the two major setbacks stated above. To be effective, this comprehensive approach clearly needs to cover all aspects of the disaster management cycle and needs to include an appropriate balance of prevention, mitigation, preparedness, response, recovery, and disaster-related development. In the past, some nations have not achieved this balance. They have, for various reasons, concentrated on post-impact relief and rehabilitation. Consequently, little or no alleviation from the impact of future disasters has been achieved. It is true that improvements can result from disasters (the "disaster-as-a-benefit syndrome"). However, this does not reduce the need for a comprehensive approach. Such an approach, because of its interrelationship with national development, is more likely to ensure that potential benefits from disaster are realized.

18. In practical disaster management terms, the overriding need is for an accurate and precise focus on the requirements at any given level of government. There is seldom scope, or indeed need, for fancy trimmings in such things as organizational structure and operational concepts. On the contrary, it is vital to define unambiguously key aspects such as:

- Possible threats,
- Resources available,
- Organizational requirements,
- Planning needs,
- Action required in relation to sectors of the disaster management cycle, and
- Training.

If this definition is correctly made and acted upon, a lean and efficient concept of countermeasures should be achievable. This should result in a system in which both government and nongovernment organizations (NGOs) are blended to provide a thoroughly professional disaster management capability.

19. In this chapter, the significance of disaster has, intentionally, been covered in limited form only. Its wider ramifications are developed in various other parts of the handbook.

*Chapter 2*

# The Disaster Threat

---

## Purpose

1. The purpose of this chapter is to outline the following:

- general effects of disaster,
- characteristics of various types of disaster,
- general countermeasures, and
- special problem areas for disaster management.

It is important for disaster managers to analyze the effects of disasters in relation to their own local circumstances. Through such analysis, it is possible to define, in advance, many of the requirements which apply to the disaster management cycle. This is especially valuable for anticipating action needed for response and recovery.

## Types of Disaster

2. The following types of disaster are covered in this chapter:

- Earthquake,
- Volcanic eruption,
- Tsunami,
- Tropical cyclone (typhoon, hurricane),
- Flood,
- Landslide,
- Bushfire (or wildfire),
- Drought,
- Epidemic,
- Major accident, and
- Civil unrest.

3. Wartime civil defense or protection measures have not been specifically included. It has been assumed that the kind of disaster management measures

advocated throughout the handbook would be broadly applicable to most wartime civil defense requirements. These measures could then be augmented or expanded by individual governments, as considered necessary.

## The General Effects of Disaster

4. Generally, typical effects of disasters may be:

- Loss of life,
- Injury,
- Damage to and destruction of property,
- Damage to and destruction of subsistence and cash crops,
- Disruption of production,
- Disruption of lifestyle,
- Loss of livelihood,
- Disruption to essential services,
- Damage to national infrastructure and disruption to governmental systems,
- National economic loss, and
- Sociological and psychological after effects.

5. Most if not all of the above effects could be expected from the types of disaster listed in the previous page.

6. In paragraphs 7–18, where the separate types of disaster are covered in more detail, some specific effects (additional to those in paragraph 4) have been included.

## Outlines of Individual Disasters

### 7. *Earthquake*

*Characteristics*

- Usually no warning. However, following a major earthquake, secondary shocks may warn of a further earthquake.
- Speed of onset usually sudden.
- Earthquake-prone areas are generally well identified and well-known.
- Major effects (see also paragraph 4) arise mainly from land movement, fracture, or slippage; specifically, they include damage (usually very severe) to structures and systems and considerable casualties due to lack of warning.

*General countermeasures*

- Developing possible warning indicators,
- Land-use regulations,
- Building regulations,
- Relocating communities, and
- Public awareness and education programs.

*Special problem areas for disaster management*

- Severe and extensive damage, creating the need for urgent countermeasures, especially search and rescue, and medical assistance;
- Difficulty of access and movement;
- Widespread loss of or damage to infrastructure, essential services, and life-support systems;
- Recovery requirements (e.g., restoration and rebuilding) may be very extensive and costly; and
- Rarity of occurrence in some areas may cause problems for economies of countermeasures and public awareness.

## 8. *Volcanic Eruption*

*Characteristics*

- Volcanoes which are likely to constitute a disaster threat are internationally well documented and, in many cases, monitored for possible activity. Usually, therefore, major eruptions can be predicted.
- Volcanic blast can destroy structures and environmental surrounds, and also cause fires, possibly including forest fires.
- Land surface cracking, resulting from volcanic explosion, may affect buildings and other structures.
- Lava flow can bury buildings and crops. It may also cause fires and render land unusable.
- Ash, in its airborne form, can affect aircraft by ingestion into engines.
- Ground deposit of ash may destroy crops and also affect land use and water supplies.
- Ash may also cause respiratory problems.
- Mud flows may arise from associated heavy rain.

*General countermeasures*

- Land-use regulations,
- Lava control systems,
- Developing a monitoring and warning system,
- Evacuation plans and arrangements,
- Relocating the population, and
- Public awareness and education programs.

*Special problems areas for disaster management*

- Access during eruption.
- Timely and accurate evacuation decision(s).
- Public apathy, especially if there is a history of false alarms or small eruptions. Thus, it may be difficult to maintain public awareness and also to implement evacuation plans.
- Control of incoming sightseers when evacuation programs are being implemented.

## 9. *Tsunami (Seismic Sea Wave)*

*Characteristics*

- The velocity of the wave depends on the depth of water where the seismic disturbance occurs. Initial wave velocity may be as high as 900 kilometer per hour (kph) (560 miles per hour [mph]), slowing to approximately 50 kph (31 mph) as the wave strikes land.
- Warning time depends on the distance from the point of wave origin.
- Speed of onset varies (see above).
- Impact on a shoreline can be preceded by a marked recession of normal water level prior to the arrival of a wave. This can result in a massive outgoing tide, followed by the incoming tsunami wave. People may be trapped when they investigate the phenomenon of the outgoing tide and then be struck by the incoming wave.
- The tsunami wave can be very destructive; wave heights of 30 meters have been known.
- Impact can cause flooding; saltwater contamination of crops, soil, and water supplies; and destruction of or damage to buildings, structures, and shoreline vegetation.

*General countermeasures*

- Optimum arrangements for receipt and dissemination of warning;
- Evacuating threatened communities from sea level/low-level areas to high ground, if sufficient warning is available;
- Land-use regulations (but these are likely to be difficult to implement if the tsunami risk is perceived as rare); and
- Public awareness and education programs.

*Special problem areas for disaster management*

- Timely dissemination of warning because of the possible short period between receipt of warning and the arrival of the tsunami wave;
- Effective evacuation time-scale;
- Search and rescue; and
- Recovery problem may be extensive and costly because of severe destruction and damage.

## 10. *Tropical Cyclone (Typhoon, Hurricane)*

*Characteristics*

- Usually long warning, derived from systematic international meteorological observation (including remote sensing);
- Speed of onset gradual;
- Tends to conform to seasonal pattern;
- Major effects arise mainly from destructive force winds, storm surge (producing inundation), and flooding from intense rainfall. Landslides may follow flooding and heavy rainfall; and
- Destruction and/or severe damage to buildings and other structures, roads, essential services, crops, and the environment generally. Major loss of life and livestock may occur.

*General countermeasures*

- Effective warning arrangements;
- Precautionary measures during warning period (e.g., boarding up buildings, closing public facilities);
- Moving people to safe shelters;
- General readiness and cleanup measures prior to an expected cyclone season (especially to reduce the risk of flying objects);
- Building regulations, and
- Public education and awareness.

*Special problem areas for disaster management*

- Assessing effects and needs may be difficult, especially due to bad weather following the impact of main disaster and to problems of access and movement caused by high damage levels;
- Widespread destruction or loss of counter-disaster resources (e.g., transport, emergency food and medical supplies, shelter materials);
- Difficulty of access and movement in carrying out urgent relief operations, especially emergency feeding, shelter and medical assistance programs;
- Search and rescue;
- Widespread destruction/disruption of essential services;
- Evacuating; and
- Rehabilitating agriculture, especially tree crops.

## 11. *Flood*

*Characteristics*

- Long, short, or no warning, depending on the type of flood (e.g., flooding within parts of a major river system may develop over a number of days or even weeks, whereas flashfloods may give no usable warning);
- Speed of onset may be gradual or sudden;
- There may be seasonal patterns to flooding; and
- Major effects arise mainly from inundation and erosion; specifically, they may include isolation of communities or areas, and involve the need for large-scale evacuation.

*General countermeasures*

- Flood control (e.g., by walls, gates, dams, dikes, and levees);
- Land-use regulations;
- Building regulations;
- Forecasting, monitoring, and warning system(s);
- Relocating population;
- Planning and arranging evacuation;
- Emergency equipment, facilities, and materials such as special floodboats, sandbags, supplies of sand, and designated volunteers who will implement emergency measures; and
- Public awareness and education programs.

*Special problem areas for disaster management*

- Difficulties of access and movement;
- Rescue;
- Medical and health difficulties (e.g., arising from sanitation problems);
- Evacuating;
- Loss of relief supplies; and
- Large-scale relief may be required until next crop harvest.

## 12. *Landslide*

*Characteristics*

- Warning period may vary. Little or no warning may be available if the cause is an earthquake. However, some general warning may be assumed in the case of landslide arising from continuous heavy rain. Minor initial landslips may give warning that heavy landslides are to follow. Natural movement of land surface can be monitored, thus providing long warning of possibility of landslides.
- Speed of onset is mostly rapid.
- Damage to structures and systems can be severe (buildings may be buried or villages swept away).
- Rivers may be blocked, causing flooding.
- Crops may be affected. Sometimes areas of crop-producing land may be lost altogether (e.g., in the major slippage of surface soils from a mountainside).
- When landslides are combined with very heavy rain and flooding, the movement of debris (e.g., remains of buildings, uprooted trees) may cause high levels of damage and destruction.

*General countermeasures*

- Land-use and building regulations;
- Monitoring systems, where applicable;
- Evacuating and/or relocating communities. Relocation has proved successful where crop-growing land areas have been lost; and
- Public awareness programs.

*Special problem areas for disaster management*

- Difficulties of access and movement in affected areas;
- Search and rescue;

- Risk of follow-up landslides may hamper response operations;
- Relocation, as distinct from temporary evacuation, may be resisted by indigenous communities;
- Rehabilitation and recovery may be complex and costly; and
- In severe cases, it may not be possible and/or cost-effective to rehabilitate the area for organized human settlement.

## 13. *Bushfire (or Wildfire)*

*Characteristics*

- Most bushfire-prone areas are well-known and well defined.
- Bushfire threat tends to be seasonal.
- Speed of onset may vary. It can be rapid under conditions of high temperatures and high wind, when major fire fronts advance very quickly. Also, fragments of fire from a main front may be carried forward by the wind, starting new fires further ahead. This is sometimes known as "spotting."
- Effects can be very destructive, especially in loss of buildings, timber, and livestock (and human life if counter-disaster arrangements are inadequate).
- Recovery from effects on the environment may take several years.
- Evacuating communities may be difficult and dangerous in the face of a major fire front.

*General countermeasures*

- Accurate risk assessment;
- Effective monitoring and warning systems, including remote sensing to define "curing" or dryingout of vegetation;
- Fire prevention regulations;
- Seasonal mitigation measures (e.g., fuel reduction);
- Building regulations; and
- Public awareness and education programs, especially to ensure that individuals, families, and communities cooperate in applying measures for prevention and mitigation, and especially that they maintain adequate standards of preparedness during the high-risk season.

*Special problem areas for disaster management*

- Maintaining adequate community awareness and preparedness;
- The arsonist problem is difficult to counter;
- Establishing and maintaining adequate firefighting resources, especially if the threat is spasmodic;
- Establishing an adequate warning system, particularly the meaning of signals (e.g., sirens) and their interpretation by threatened communities;
- Timely dissemination of warning and, if applicable, decision to evacuate;
- Long-term recovery may be prolonged due to high levels of environmental damage and destruction; and
- Evacuation movements, either out of affected areas or to safe havens within such areas.

## 14. *Drought*

*Characteristics*

- Major areas liable to drought are usually well-known;
- Periods of drought can be prolonged;
- Area(s) affected may be very large;
- Long warning;
- Effects on agriculture, livestock, rural industry production, and human habitation may be severe. This may lead to prolonged food shortages or famine;
- Long-term effects can be in the form of severe economic loss, erosion which affects future habitation and production, and sometimes abandonment of large tracts of land;
- Man-made activities may aggravate the possibility and extent of the drought problem (e.g., overgrazing of agricultural land, destruction of forests or similar areas); and
- The inability and/or unwillingness of the population to move from drought-prone areas may exacerbate the problem.

*General countermeasures*

- There are few, if any, quick and easy solutions to the drought problem; effective countermeasures tend to be mostly long term;
- The long-term resolution of drought problems usually rests with national governments and involves major policy decisions;

- Since these decisions involve human settlement, they are often sensitive and difficult ones;
- International cooperation and assistance usually play an important part in coping with major drought problems;
- Land management and special plans (e.g., for irrigation);
- Response to drought-caused emergencies usually includes providing food and water supply, medical and health assistance (including monitoring of sanitation and possibility of epidemic), and emergency accommodation (may be on an organized camp or similar basis); and
- Information programs, especially to assist aspects such as land management.

*Special problem areas for disaster management*

- Response requirements (e.g., feeding programs) may be extensive and prolonged, thus involving major commitment and expenditure of resources.
- Prolonged drought may undermine self-reliance of affected communities, thus making it difficult to withdraw disaster management assistance.
- Logistic requirements may exceed in-country capability, particularly if large inputs of outside (international) commodities are involved.

## 15. *Epidemic*

*Characteristics*

- Disaster-related epidemic arises generally from the disrupted living conditions which follow disaster impact.
- Epidemic may arise from:
  food sources;
  water sources;
  inadequate medical and health facilities/standards;
  malnutrition; and
  vector-borne sources (e.g., mosquitoes).
- Types of disease include:
  hepatitis,
  typhoid,
  diptheria,
  malaria,
  cholera,
  influenza,
  enteritis,

diarrhea,
skin diseases, and
food poisoning.

- Under post-impact conditions, when personnel and facilities may be limited, outbreaks may prove difficult to contain and control. This may particularly apply if community health education is substandard.
- Warning (i.e., risk) is self-evident in most post-impact circumstances.
- Speed of onset is mostly rapid.

*General countermeasures*

- An effective medical and health sub-plan within the overall local or area counter-disaster plan. This medical and health plan needs particularly to cover preparedness measures and the capability to deal with post-disaster eventualities;
- Close post-disaster monitoring of medical and health aspects;
- Reinforcement of medical resources and supplies in anticipation of epidemic outbreak; and
- Public awareness and education, both before and after disaster impact.

*Special problem areas for disaster management*

- Loss of medical and health resources (e.g., clinics, medical supplies) during disaster impact (e.g., by a cyclone) may inhibit response capability;
- In-country shortage of special equipment (e.g., water purifying plant).
- Integrating outside (international) medical and health assistance with local systems; and
- Containing and controlling common diseases (e.g., enteritis and diarrhea) which can have a mass effect, especially if relevant medical and health resources are severely limited.

## 16. *Major Accident*

*Characteristics*

- Usually violent in nature (e.g., industrial or other explosion, aircraft crash, major fire, train collision);
- Can have limited or widespread effect (e.g., an aircraft crash may affect only those on board, whereas an explosion involving hazardous chemicals may affect a wide area of the population);

- Mostly limited or no warning, though there may be longer warning of effects of, say, chemical or oil spill; and
- Speed of onset usually rapid.

*General countermeasures*

- Good physical planning (e.g., the siting of potentially accident-prone buildings or complexes);
- Special building regulations, if applicable;
- Good in-house safety and management standards/procedures, including evacuation plans and periodic tests;
- Effective organizational emergency services (e.g., fire services and rescue teams) which are available to immediately respond prior to the arrival of public emergency services;
- Effective community or area disaster plans so that coordinated response can be achieved; and
- Training in handling the effects of specific hazards.

*Special problem areas for disaster management*

- Unexpected nature of accidents may pose problems of reaction and response time;
- Response problems may be severe, extensive, and difficult (e.g., rescue from a building collapse, or in circumstances where a chemical or radiation hazard exists, or where there are multiple casualties such as in a major rail accident); and
- Identifying victim may be difficult in some cases.

## 17. *Civil Unrest*

*Characteristics*

- Usually the responsibility of police, paramilitary, and armed forces. However, other emergency services such as fire services, medical authorities, and welfare agencies become involved;
- Violent and disruptive activities occur (e.g., bombing, armed clashes, mob demonstrations, and violence);
- Patterns of civil unrest are difficult to predict. Therefore, effective warning may also be difficult;
- In many civil unrest circumstances, especially terrorism, the instigators have the initiative, thus complicating the task of law enforcement authorities.

*General countermeasures*

- Firmly applying law and order regulations and requirements;
- Imposing special emergency measures and regulations (e.g., restricted movement, curfews, and security checks); and
- Positive information programs aimed at maintaining majority public support for government action against disruptive elements/factions.

*Special problem areas for disaster management*

- Overloading of resource organizations (e.g., medical authorities, welfare agencies, and essential services) because of demands of civil unrest incidents, in addition to normal commitments; and
- Difficulty of integrating "peacetime" resource organizations (noncombatant in nature) with "military type" operations which are necessary to deal with violent civil unrest.

### 18. *Other Disasters*

In some countries, disasters in addition to those listed in paragraph 2 of this chapter may apply (e.g., severe outbreaks of animal diseases that threaten rural industries, food production, and so on). In such cases, it is suggested that disaster managers should assess the particular disaster(s) along the above lines.

## Process of Defining the Disaster Threat

19. There can be variations in the process by which disaster management authorities and associated scientific and technical agencies define the threat from any particular form of disaster. Also, the capability to define disaster threats accurately is likely to vary between different countries. This depends on the standards of disaster management and other disaster-related activities, including study and research.

20. However, a basic theme or pattern comprises the following main areas of action:

- Identifying the hazards;
- Assessing the vulnerability of settlements, communities, and assets to relevant hazards; and
- Evaluating the risks.

## 21. *Identifying Hazards*

A hazard can be generally described as a threatening event. It may take the form of a natural phenomenon, such as a possible cyclone, or it may be basically artificial, such as the accidental release of a hazardous substance from an industrial complex. The process of identifying hazards obviously involves carefully surveying the country or region concerned. This survey may require inputs from a variety of specialist agencies and authorities, including information on past disaster-related events. Usually, this identification process includes hazard mapping, which establishes geographically where natural and artificial hazards may occur. The relationship of these hazards to human settlements and institutions then provides a valuable indication of the risks that may be involved.

## 22. *Assessing Vulnerability*

For the foregoing identification of hazards, it becomes possible to identify—with reasonable accuracy—those settlements, communities, and assets that are especially vulnerable to disaster-caused damage or destruction.

## 23. *Evaluating Risk*

Risk has two dimensions, frequency and magnitude/intensity. Evaluating risk is done by relating a natural or artificial hazard to the primary characteristics (e.g., population distribution and development aspects) and vulnerability of the area concerned. This process particularly identifies high-risk areas and is the basis for producing risk maps. For example, risk mapping of a bushfire-prone area would indicate the likelihood of fires occurring and the degree to which those fires would affect communities within the area. Similarly, for a flood-prone area, risk mapping would show the likely levels of inundation for various flood intensities.

## Use of Disaster Threat Information

24. The hazard, vulnerability, and risk information in paragraphs 19–23 above is, of course, only an outline of what is a detailed and extensive procedure. However, it serves to illustrate the value of disaster threat information, as applied to practical disaster management. For instance, it is suggested that the information in this chapter should be used and, indeed, is essential for the following:

- The formulation of disaster plans, especially the measures within such plans that deal with preparedness, response, and recovery.

- The formulation of relevant programs for disaster-related training and public awareness;
- The definition and application of measures that can reduce vulnerability in specific cases/areas; and
- Formulation and use of long-term programs of mitigation and prevention.

## Further Information

25. Further relevant information is contained in Appendix A which deals with vulnerability analysis. However, as indicated in paragraph 19, there can be variations in the process of defining threats. Therefore, users of this handbook should consider local circumstances and factors when using the process.

*Chapter 3*

# National Disaster Management Policy

---

## Purpose

1. The purpose of this chapter is to outline the following:

- need to clearly define a national disaster management policy;
- process for defining this policy;
- main elements of such a policy; and
- arrangements for monitoring and reviewing the national policy

## Clear Definition of National Policy

2. Clear definition of national disaster management policy is essential if a country is to establish and maintain adequate arrangements to deal with all aspects of its disaster threat. This applies to all levels of the national structure and organization—that is, from the national government to the local government or community level. If such a policy does not exist, arrangements to deal with disaster will be ill-defined and inadequate. Consequently, loss of material and human resources will arise; the nation, as a whole, will suffer.

3. A strong and clear policy offers most, if not all, of the following advantages:

- demonstrated lead from government in disaster-related affairs;
- foundation for appropriate legislation and associated regulations;
- basis for sound organization and clear allocation of responsibilities;
- overall direction for ensuring optimum use of resources against a carefully assessed threat; and
- national competence and self-reliance that is likely to engender optimum international assistance when the need arises.

## The Process of Policy Definition

4. To define a national disaster management policy, it is necessary to consider certain main factors or pillars. For most countries, the following will usually apply:

- defining accurately the disaster threat;
- identifying the effects which are likely to be caused by the threat;
- assessing the resources available to deal with the threat;
- organizational arrangements which are required to prepare for, respond to, and recover from disaster events;
- defining how a national disaster management policy interlocks with other aspects of national policy, especially those concerned with national development and protection of the environment;
- any other specific national factors which may be applicable.

### 5. *The Disaster Threat and Likely Effects*

Information on the disaster threat and on the effects likely to arise from disaster is contained in Chapter 2. However, for purposes of defining national policy, these two aspects would need to be carefully reviewed. This is necessary to ensure that a correct relationship is established between the threat and its effects on the one hand, and the policy itself on the other hand. This is particularly relevant to the selection of policy options (paragraphs 10–11) and to the priorities within the national policy which are given to these options. For instance, from a purely disaster management viewpoint, the most effective countermeasure to a flood threat may be a policy of prevention based on an extensive flood control system. However, when it comes to national policy, preventive measures of this kind may not be financially feasible.

### 6. *Assessing Available Resources*

Obviously, it is no use framing a national disaster management policy which is beyond the capacity of available resources. The reverse sense applies; that is, national policy has to be balanced with the various existing resources in terms of equipment, facilities, and personnel. In assessing resources, it is essential to consider the widest range of both government organizations and NGOs. It is also reasonable to take into account resources which are likely to be forthcoming from international sources. Further information concerning counter-disaster resources is contained in Chapter 8.

## 7. *Organizational Arrangements*

It is generally accepted that the primary responsibility for dealing with disaster must rest with national government. In addition, an important disaster management concept is to ensure optimum use of existing resources, the majority of which tend to be under government direction. It follows, therefore, that the organizational arrangements needed for dealing with disaster (both before, during, and after a disaster) are best based on the government structure. Indeed, experience has shown that it is neither wise nor effective to try to switch to some special organizational arrangements for disaster purposes.

Most countries, therefore, use their existing government structure as the basis for dealing with disaster. They then augment this by establishing such specialized agencies or sections as may be deemed necessary; for example, a National Disaster Council (NDC) for policy purposes, a National Disaster Management Office (NDMO) for day-to-day working affairs, and appropriate sections at lower levels of government.

Thus, in framing a national disaster management policy, these organizational aspects must be carefully considered and included in the relevant policy statement. In this regard, it is usually prudent to frame the policy so that minor adjustments to organizational arrangements can be made without having to amend the policy itself.

## 8. *Interrelation of National Disaster Management Policy with Other National Policies*

It is well understood and recognized that governments in most countries are, of necessity, concerned with a whole range of major policy areas. These usually include economic and social development, health, education, and so on. Such major policy areas will, necessarily, be prioritized from a government standpoint. It is unrealistic, therefore, to expect that disaster management policy will be prioritized, for instance, in allocating funding and resources that cannot be positively justified.

Therefore, those responsible for drafting and formulating a national disaster policy must aim to achieve an appropriate balance and interrelationship with other national policies. This obviously involves carefully considering these other policies, especially with a view to ensuring as far as possible the compatibility of interest. Indeed, in some cases, it may be that an astute disaster management policy can offer advantages to other national policies. Where this is so, mutual support and the advantages that this brings is likely forthcoming.

Two areas of government policy tend to have mutual interests with disaster management. They are:

- *National development*– Disaster and national development are, in reality, closely related. This especially applies if the disaster threat is significant. For instance, national development planning needs to consider the possible effects that disaster may have on the various programs and projects involved. In turn, however, such programs and projects may affect the nation's ability to cope with disaster because while some of them may reduce risk and vulnerability, others can actually increase it. In addition, disaster events often open subsequent possibilities of improving various aspects of progress and development. This is known as the "disaster-as-a-benefit syndrome," as already mentioned in Chapter 1. These improvements and benefits may be possible in various fields of activity; for instance, building standards, transport systems, town planning, and so on.

- *Protection of the environment* – Widespread international concern has prompted most national governments to direct particular attention toward protecting the environment. Since much disaster management activity is concerned with environmental aspects (e.g., flood, drought, cyclone) it is sensible to maintain close collaboration between disaster management and environmental policies. Indeed, some countries are already moving toward a single integrated policy for disaster and the environment.

It is suggested that these two policy areas illustrate the need for and benefit from a disaster management policy that is appropriately interrelated with other policies. Disaster management officials, especially at national government level, would therefore be well advised to regard this concept as being of key importance whenever disaster policy is being formulated or reviewed.

## 9. *Other Factors*

It may be that—in addition to the factors discussed in paragraphs 4–8 above—others are relevant to individual country circumstances. Therefore, before embarking on the drafting of a national policy, disaster management officials should conduct a careful survey to identify any such additional factors.

## Main Elements of National Policy

### 10. *Options Available*

In identifying what options are available as elements of national policy, it is useful to study the disaster management cycle, which is discussed in the introductory notes and Chapter 5. From these sources, it will be readily seen that all segments of the cycle are possible elements, as follows: prevention, mitigation, preparedness, response, recovery, and development.

These segments should be scrutinized against the following definitions (or alternative definitions if the latter appear more suitable for a particular national policy).

- *Prevention* – Measures aimed at impeding the occurrence of a disaster and/or preventing such an occurrence having harmful effects on communities. Constructing a dam or levee to control floodwaters is an example of a preventive measure. Controlled burning off in a bushfire-prone area prior to the high-risk season is another example.

- *Mitigation* – Action taken (usually in the form of specific programs) to reduce the effects of a disaster on a nation or community. For instance, developing and applying building codes can reduce damage and loss in the event of earthquakes and cyclones. The term normally implies that while it may be possible to prevent some disaster effects, other effects will persist and can be modified or reduced if appropriate action is taken.

- *Preparedness* – Measures which enable governments, communities, and individuals to respond rapidly and effectively to disaster situations. Preparedness measures include formulating viable counter-disaster plans, maintaining inventories of resources, and training of personnel.

- *Response* – Response measures are usually those taken immediately prior to and following disaster impact. They are directed toward saving life, protecting property, and dealing with the immediate damage and other effects caused by the disaster.

- *Recovery* – The process by which communities and the nation are assisted in returning to their proper levels of functioning following a disaster. The recovery process can be very protracted, taking 5–10 years or even more. Recovery is usually taken as including other aspects such as restoration and reconstruction.

- *Development* – The progressive advancement and modernization of societies, in this case as it interrelates with the effects of disaster and with disaster management.

## 11. *Choice of Options*

It is clear from the foregoing definitions that there is little real option or choice in the case of preparedness, response, and recovery, and that these must be covered in national policy to the best possible extent. Thus, they constitute option priorities 1–3 of a national policy. In fact, some countries —initially, at any rate—have found it necessary to limit their national policy to these three options, largely because of financial considerations.

The other aspects are best considered and selected against the following criteria, and probably in the order of option shown below. However, it is emphasized that individual national circumstances might require different priorities from the ones suggested. For instance, it might be highly important to implement a specific flood prevention program to protect a vital national production or other asset.

- *Development – Option Priority 4*
  As stated in paragraph 8, there is a clear and valid interrelationship between disaster and national development and, in particular, each can affect the other. Also, a post-disaster period usually offers the opportunity to enhance aspects of national development.

- *Mitigation – Option Priority 5*
  If it is accepted, as inferred in the definition contained in paragraph 10 that not all disaster effects may be prevented, but that these effects can be mitigated, then mitigation logically becomes the next option. Its particular advantage, in terms of national policy, is that it offers the possibility of reducing damage and loss, thus easing the burden on the government and the nation.

- *Prevention – Option Priority 6*
  As inferred at the beginning of this paragraph, adopting programs of prevention may be problematical. In some cases, because of cost, they may have to be omitted or delayed. In other cases, they may be mandatory to protect a specific vital interest. The option priority is

therefore best decided on an individual national basis. Thus, it may assist national policy makers to use the combined category heading of prevention/mitigation.

### 12. *A Possible Policy Format*

In summarizing what has been said concerning the main elements of national policy, a possible policy is given on the following page. However, the method by which the policy is issued is a matter for individual national choice. For instance, it might be issued by:

- government decree,
- government statement,
- legislation,
- regulation, and
- other means.

## Arrangements for Monitoring and Review

13. The policy statement itself should designate responsibility for monitoring and reviewing the national policy statement. Usually, it is appropriate to allocate this responsibility to the responsible minister (or ministerial body), who could normally be expected to delegate it to the senior disaster management body (e.g., NDC).

14. Post-disaster review (Chapter 20) will often provide a good indication of whether existing national policy is valid, or whether it should be reviewed and amended.

## SUGGESTED STATEMENT OF NATIONAL DISASTER MANAGEMENT POLICY

Introduction

Including summary of the disaster threat

Purpose of the Policy Statement

The purpose of this statement is to define the national policy for dealing with the disaster threat to Exland

or

The purpose of this statement is to define the national policy for disaster management necessary to counter the disaster threat to Exland.

The Overall National Aim

The overall national aim regarding disaster management is to establish and maintain the optimum system for dealing with disaster which is achievable within national resources.

Responsibilities for Disaster Management

at national level (e.g., ministerial responsibility, departmental responsibilities)
at other levels

Main Elements of Disaster Management

Prevention,
Mitigation,
Preparedness (with definitions),
Response,
Recovery, and
Development

Priorities Accorded to Main Elements

Planning

National Disaster Plan
Other Plans

Organization

Use of existing government structure
Arrangements for augmenting government structure (e.g., NDC)

Important Disaster Management Concepts

Optimum coordination and use of resources
Community involvement and awareness to develop preparedness and self-reliance.

Legislation

Special disaster legislation
Other emergency powers

Relationship with other National Policies

National development,
Protection of environment, and
Others

Other Sections as Required

e.g., financial arrangements

Viability of National Policy

Responsibility for monitoring and review

Government Authorization

This policy is issued under Government Decree (or Cabinet decision) No., dated

*Chapter 4*

# Major Requirements for Coping with Disaster

## Purpose

1. The purpose of this chapter is to outline the major requirements for coping with disaster, especially in terms of:

- organization,
- planning,
- use of resources,
- specialist skills, and
- training needs.

The chapter is intended to provide a preliminary ready reference to major areas of disaster management, especially for persons who may be initially assuming duties in this field. These areas are expounded in later chapters and cross-reference to these chapters has been made.

## Scope of Disaster-Related Action

2. In assessing the scale of organizational, planning, and other measures which need to be undertaken to deal with disaster, it is useful to bear in mind the range of responsibilities and tasks that are likely to arise. For instance:

- providing and disseminating warning;
- search and rescue;
- surveying, assessing, and reporting disaster effects;
- treating and caring of victims;
- clearing debris and rehabilitating roads, airfields, railway systems, ports, and other key areas;
- providing emergency food and water supplies;
- providing shelter;
- evacuating individuals, groups, and communities, and livestock;

- providing health and sanitation measures;
- restoring essential services such as communications, water and power supply;
- directing and coordinating counter-disaster measures;
- informing and advising the public;
- financially assisting victims immediately;
- maintaining public morale;
- counseling victims and relatives;
- controlling and distributing emergency supplies;
- liaising with the media;
- rehabilitating crops, production, and other aspects of subsistence and livelihood;
- emergency building programs;
- measures for long-term recovery; and
- applying emergency regulations.

3. The scope and variety of these tasks underline the need for a carefully organized and systematic approach to all aspects of disaster management. Ad hoc measures, even if based on long experience and traditional action, are likely inadequate, although obviously they can make a useful contribution.

## The Philosophy for Coping with Disaster

4. Before the requirements for coping with disaster can be determined and adequately met, it is essential to bear in mind the simple but nonetheless important philosophy which is involved. This is that disaster can have wide-ranging effects on a country, its government, and its people. Therefore, the primary responsibility for coping with disaster must rest with government. Government is responsible to the people for meeting the needs created by disaster, in the same way that it is responsible for other aspects of national life. However, a government may, and usually does, need help from NGOs, the private sector, general public and, in many cases, from its international development partners also.

5. This being so, it is government organization and resources that necessarily have to bear the brunt of counter-disaster action. Further, when government departments and agencies (including emergency services) have to deal with disaster, they invariably have to accept a considerable additional workload. Moreover, they usually have to function under pressure and in cooperation with a variety of other agencies (government or otherwise). A further point is that disaster produces some needs that may not be covered by normal government organization such as mass feeding.

6. This is why, in essence, a disaster management system is needed. But —and this is important—the disaster management system does not control other government departments; it does not tell those departments how to do their jobs. The main function of the disaster management system is to ensure that at all times, and as far as possible, the resources and operations of these government departments and, as appropriate, NGOs are coordinated to produce the best possible counter-disaster effort.

7. In sum, therefore, the simple philosophy for coping with disaster is one of government and people working together in a coordinated way via a coherent disaster management system. The remainder of this chapter should be considered against this background.

## Foundation Factors

8. Before the major requirements for coping with disaster can be established, it is essential that certain basic or foundation factors should be effectively covered. These include:

- an understanding of the significance of disaster, especially at key levels throughout the government structure;
- a clear assessment of the disaster threat; and
- an effective disaster management policy.

9. These factors have been considered in Chapters 1–3. However, their overall importance is reemphasized here because they are fundamental to the requirements of organization, planning, resource utilization, and training which are discussed in the remainder of this chapter.

## Organization

10. Some people have difficulty in understanding what exactly comprises a disaster management organization or system. It may be useful, therefore, to state some things that such an organization does not do.

- It does not duplicate normal government organization.
- It does not act independently of government.
- It does not control other government departments.
- It does not act outside the charter or terms of reference given to it by government, except perhaps in cases of extreme urgency.

These things being so—and they apply in most countries—it becomes clear that when we talk about disaster management organization, or counter-disaster organization, we are really talking about an organizational system. Moreover, this is a system which is essentially an adjunct to the normal government system, designed to enable the latter to deal effectively with the special demands which may arise from disaster. The extent of these demands is well illustrated by paragraph 2.

11. Consideration of a simple example helps illustrate the various foregoing points. In the following diagram, for ease of illustration, it has been assumed that national policy—perhaps for financial reason—has been confined primarily to preparedness, response, and recovery. However, if other aspects had been included, they could easily be incorporated into the system shown. It is emphasized, of course, that the system illustrated is only one of several variations which could be used.

## 12. *Important Organizational Considerations*

A more detailed coverage of organizational aspects is contained in Chapter 11. That chapter provides details, for instance, of the composition of an operations control group (OCG), the functions, management, and staffing of a national emergency operations center, and so on.

However, the following certain important organizational considerations are worth noting at this stage:

- *The nature of the system*
  As emphasized in paragraph 10, the system is not some form of specialized setup superimposed on the existing government structure. On the contrary, it adapts existing resources for disaster purposes. For instance, an NDC would not be made up of special disaster management officials. It would normally comprise senior government officials (e.g., heads of departments) and representatives of NGOs because these people best understand and are responsible for the availability and utilization of resources in their respective spheres. They also understand the management factors and systems which apply within these two spheres. Obviously, members of the NDC would be required to have knowledge of disaster management, acquired through training and experience.

- *The utilization of the total government structure*
  The system utilizes the total extent and depth of an existing government structure; that is, from a national level through provincial (or its equivalent) level, down to local government and community level. This government structure is, by its nature, permanent. It therefore makes sense to utilize it for all disaster-related purposes. In any event, experience indicates that it is unwise to try to switch to some alternative system or structure purely to deal with disaster.

- *Coordination of nongovernment resources*
  Nongovernment resources can be easily coordinated into the system. This can be done, for instance, through the NDC—as already mentioned—through the NDMO (with which there can be an NGO liaison link) and through the OCG and the technical advisory team (both of which there can be NGO representation).

- *Community involvement*
  Disasters happen at community level. They may affect several communities at one and the same time, but these communities that constitute what might be called the "disaster front." Any disaster management organization must therefore cater to appropriate community involvement and participation. This applies particularly to relevant aspects of preparedness, response, and recovery action. In fact, in framing and/or checking organizational effectiveness, it is advisable to consider requirements at community level first and then, subsequently, to tailor arrangements at higher levels so that these fit requirements at the "disaster front."

- *Specialist facilities and systems*
  A disaster management organizational system is essentially a dynamic entity. All phases of the disaster management cycle infer and involve action. This obviously requires a range of specialist facilities and systems, usually needed to cover things such as:
  - direction and coordination of disaster-related action;
  - emergency operations center activities;
  - alerting and activating the disaster management system, as and when required;
  - communications;
  - warning;

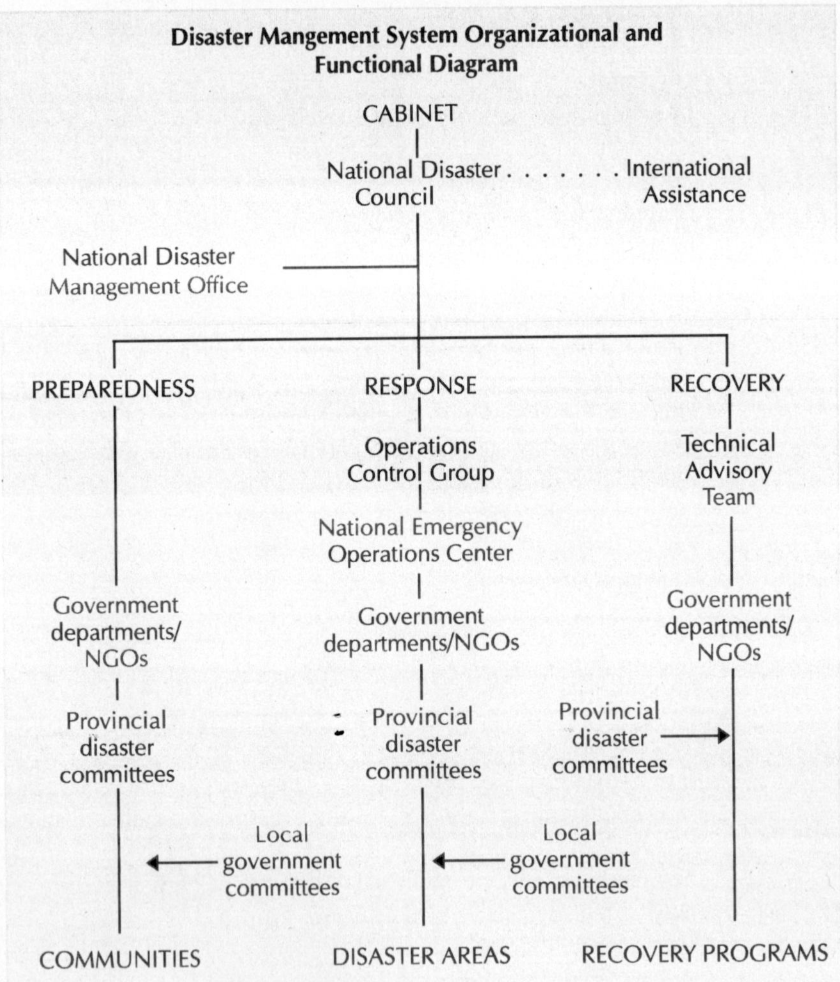


- survey and assessment;
- information management; and
- emergency logistics.

In considering overall organizational requirements, therefore, it is essential to determine the range of specialist facilities and systems which will apply to any particular set of circumstances.

## 13. *An Important Product*

Clearly, a sound organizational system is one major key to successful disaster management. However, in considering and applying what has been said in paragraphs 10–12 above, it should be noted that effective organization produces one particularly important product. This is what is sometimes called "operational coherency" during response operations. Operational coherency means that the disaster direction/coordination authority—and therefore the government—has a clear picture at all times of the needs caused by the disaster and the pattern of action being taken to cope with such needs. If operational coherency is lost—even for short periods—response operations can be adversely affected, sometimes very seriously.

### 14. *Further Information on Organization*

Further information on organizational aspects is contained in Chapter 11.

## Planning

### 15. *The Need for Counter-Disaster Plans*

The need for counter-disaster plans is sometimes questioned. Proponents of this view take the attitude that disaster will occur whether we like it or not. Therefore, it is better to let nature take its course, then use all available means of assistance to build a better tomorrow. However, the vast majority of international experience indicates that where plans did not exist, or where planning was inadequate, then the effects of disaster on countries and their peoples have been worse than would have otherwise been the case. A senior disaster management official once wrote to the author of this handbook as follows: "When we look back on the cyclone and what it did to our country, there is no doubt that hundreds of people are alive here today because we had proper disaster plans."

There can be little doubt that an effective basis of planning and the maintenance of relevant plans does offer the following advantages:

- clear and coherent approach to dealing with disaster;
- common reference for all departments and authorities which have roles in counter-disaster activity;
- basis for coordinated action;
- clear allocation of responsibilities;
- focus for disaster-related training; and
- setting against which to review and evaluate current and future disaster management requirements.

### 16. *Pitfalls in Planning*

There are, of course, pitfalls in the planning field. Plans must be reviewed and revised as necessary so that they are fully updated. Indeed, it is true that an unrevised and outmoded plan may cause more trouble than no plan at all.

There is also a serious danger from nominalism in planning. Nominalism is often regarded as the deadly sin of a disaster manager. It is relatively easy—given a specific set of circumstances—to produce a counter-disaster plan, put it in an attractive cover, and circulate it to all concerned. However, this can amount to little or nothing more than a useless front or facade. Because if the system which is necessary to implement the plan is not fully effective, then, quite simply, the plan will not work. In other words, if there is lack of necessary funding, personnel, equipment, facilities, systems, training support and so on, the plan becomes nominalistic; that is, a plan in name only. This has occurred in some countries, usually because of:

- lack of a clear national policy,
- inadequate funding,
- limited expertise, or
- other similar reasons.

So what is very much needed in planning is realism, not nominalism.

Another planning pitfall concerns what is often called "moving the goalposts." This usually takes the form of major changes to government policy, structure, or organization. The result is that the plan no longer fits the realities of the situation that must therefore be amended; otherwise, serious problems will arise. The sequence of events in a certain disaster-prone country illustrates this point.

| | | |
|---|---|---|
| Year X | – | The country was hit by a disaster, resulting in great damage, destruction, and loss. |
| Year X + 1 | – | The government, with help from the United Nations, produced a national disaster plan. |
| Year X + 3 | – | Another disaster struck the country, the plan was brought into action and the situation handled satisfactorily.<br>The plan was subjected to post-disaster review and, apart from minor amendments, was confirmed. |
| Year X + 7 | – | A further disaster occurred and the plan was brought into action. This time, the plan did not work effectively. |

Subsequently, it became clear that there were two major reasons for the apparently proven plan not working in year X + 7:

i. The plan had not been reviewed during the previous 4 years and, therefore, became stale.
ii. In the same 4-year period, a new government had made significant changes to the government structure. Thus, the plan was no longer applicable to current circumstances.

### 17. *Other Critical Areas in Planning*

Several other areas of planning can prove critical when plans are implemented in response to a disaster event. They include the following:

- crisis pressure which arises from disaster impact;
- effect of disaster impact on the organizational structure;
- deficiencies in the information management system;
- inadequacy of planning for post-impact survey, assessment, and reporting;
- allocation of tasks to resource organizations; and
- coordination of counter-disaster effort.

These critical areas are covered in later chapters.

### 18. *Further Information on Planning*

Further information on counter-disaster plans, including planning guidelines, is contained in Chapter 12.

## Use of Resources

19. One most difficult problem in disaster management is to achieve the optimum use of available resources. Difficulties tend to arise for some or all of the following reasons:

- counter-disaster planning has been inadequate;
- all potential resources have not been identified during the planning process;
- resource organizations have not been accurately assessed in terms of their capability, availability, and durability;
- allocating tasks to resource organizations has been inaccurate or inappropriate;
- resource organizations are not sufficiently skilled in carrying out allotted tasks;
- some resources may be destroyed or put out of action by disaster impact;

- delays may occur in the availability and application of resources, also due to disaster impact;
- resource organizations have had insufficient practice in their disaster roles, especially in coordinating their activities with those of other resource organizations;
- system for directing the use of resource organizations, especially for coordinating their activities, is inadequate;
- inadequate and/or inaccurate information may lead to ineffective deployment and use of resources; and
- poor direction of resources may result in duplication or gaps in disaster-response activities.

## 20. *An Important Management Proviso*

The resources which are used to deal with disaster situations tend to be a mixture—sometimes a complex one—of:

- government departments and agencies (sometimes including military forces);
- NGOs;
- community individuals and groups; and
- international teams or volunteers.

During the planning process, it may not have been possible to accurately envisage the nature of some of these resources: for example, international volunteer teams. Therefore, the direction/coordination authority in charge of response operations may face problems in optimally utilizing all resources.

An important management proviso in all these circumstances is to correctly use the management systems within individual resource organizations. Certainly, the direction/coordination authority should, for example, tell a public works department what tasks need to be carried out such as for debris clearance, opening of roads, etc.) and what priorities should be given to these tasks. However, the direction/coordination authority should not try to tell public works how to carry out such tasks; this is essentially a matter for the department's own management.

If this management proviso is strictly observed, then:

- there will be no confusion between the management role of the direction/coordination authority and that of the resource organization;
- there is unlikely to be conflict of management;

- tasks will be carried out accurately and efficiently; and
- overall effectiveness of response operations will be maintained at optimum level.

21. ***Important Resource Management Needs***

Given that the foregoing important proviso is recognized and applied, resource management—and thus effective utilization—depends largely on four major needs, as follows:

- a capable emergency operations center (EOC) system;
- a good information picture;
- effective communication between the direction/coordination authority and individual resource organizations; and
- sensible commitment of resource organizations to operational tasks, bearing in mind their capability and durability.

22. ***Further Information on Resources***

Further information concerning resources and resource organizations is given in Chapter 8.

## Specialist Skills

23. In paragraphs 4–21, various aspects of the following subjects have been covered:

- the philosophy for coping with disaster,
- foundation factors,
- organization,
- planning, and
- use of resources.

Throughout these aspects, a common theme is apparent. It is that in dealing with disaster, every effort needs to be made to extract maximum effectiveness from existing systems and resources.

24. However, the scope of disaster-related action shown in paragraph 2 above indicates that a multiplicity of skills is necessary if disaster management is to be effective. While most of these skills are likely to exist within various organizations and the community, they may be insufficient in strength and numbers to cope with disaster, especially on a large scale. This may particularly apply to the following:

- search and rescue;
- survey, assessment, and reporting;
- first aid;
- mobile medical teams;
- evacuation;
- emergency welfare (e.g., mass feeding programs);
- emergency shelter (e.g., erection of tentage, emergency building repairs);
- emergency logistics;
- staffing of EOCs, including mobile ones; and
- information management.

In this connection, the need for developing the leadership skills of team leaders/managers should be borne in mind.

25. It is suggested, therefore, that a broad assessment of specialist skills requirements should be made and compared against the number of skilled personnel available. In this way, the needs for training in specialist skills can be established broadly.

## Training Needs

### 26. *The Importance of Trained Personnel*

Moving into the more general training field, it is self-evident that trained personnel constitute a key component in effective disaster management. By contrast, unskilled and untrained disaster operatives can well be a menace to themselves and to other people, as the following examples indicate:

- Following an explosion in a building, the situation was made more difficult and complex than it was originally by unskilled people attempting to save lives. There was no clearly identifiable search and rescue coordinator on-site and this—coupled with the unskilled effort—resulted in an unsystematic search and rescue pattern. For, instance, when some would-be rescuers moved planks, bricks, and other debris to one part of the rescue site, others almost immediately moved them back again. The resultant inefficiency of the total search and rescue effort was blatantly obvious.
- An unskilled person attempting to fight a serious outbreak of fire as part of a volunteer team, almost immediately became a casualty himself. Not realizing the fire-prone nature of the synthetic clothing that he was wearing, he himself was set afire. Thus, not only did the

volunteer team lose a valuable member, it also had a casualty who had to be given urgent first-aid attention.

- A wrong decision—obviously based on lack of disaster management knowledge and skill—resulted in a large number of people being sheltered from a cyclone in buildings that were structurally inadequate. More than 40 people died when the buildings collapsed.

Unfortunately, many other similar examples serve as reminders of the dangers arising from inadequate knowledge and training.

## 27. *Major Training Considerations*

International experience indicates that important policy considerations need to be applied to disaster management training. They are as follows:

- Nations should identify their own training needs and frame their individual policies accordingly. To take examples from other countries or academic sources, and try to apply them as exact "blueprints," is unlikely to prove satisfactory.
- Training programs must be designed to be compatible with and support disaster plans.
- Responsibility for training must be clearly defined in national policy statements and counter-disaster plans.

## 28. *Scope of Training Activity*

One important first step in establishing training policies and programs is to identify clearly the scope of training activity required. This is best done by closely examining the following:

- national disaster management policy;
- the overall disaster management structure, through all levels;
- all relevant plans including, for instance, those of individual government departments and NGOs;
- requirements for public education and awareness with which training programs need to be compatible; and
- other relevant sources of information (e.g., comparison with overseas training programs).

## 29. *Training Policy*

Once the scope of necessary training activity has been defined, it should be translated into a training policy. This is best done in the form of a simple policy document that clearly sets out the following:

- major training consideration (see paragraph 27);
- aim of training policy;
- responsibility for implementing a training policy;
- scope of training to be carried out (e.g., disaster management areas to be covered and levels at which programs will be implemented;
- process for reviewing and updating training programs; and
- periodic issue (annually or otherwise) of training programs.

In framing and implementing a training policy, it must be borne in mind that training is a dynamic activity which is susceptible to changing requirements. For instance, in a given training program, the need for some activities will diminish, while the need for other activities will arise. Consequently, assessing the training pattern annually is usually worth carrying out so that programs are kept compatible with realistic training needs.

## 30. *Implementation of Training*

For most countries, it is appropriate to implement a twofold training, as follows:

- *In-countryt raining*
  This should essentially address domestic training needs in accordance with national policy and its selected programs. Such training should use in-country resources to the best possible extent. It should usually comprise a variety of activities such as seminars, workshops, courses, exercises, etc. (Chapter 21).

- *International training*
  International training is usually designed to broaden the knowledge of key disaster management officials, mainly through the interchange of views, ideas, and experience which it provides. It should be regarded as an important adjunct to in-country training but not as a substitute for the latter. (Chapter 21).

## 31. *Further Information on Training*

The points covered in paragraphs 26–30 are intended to underline certain major training issues only. Further detailed information is given in Chapter 21.

*Chapter 5*

# The Disaster Management Cycle

## Purpose

1. The purpose of this chapter is to outline the main considerations concerning the cycle of events that applies to disaster management, including the following:

- disaster management cycle used in this handbook,
- composition of the main activities within the cycle, and
- application of the cycle to practical disaster management.

## The Disaster Management Cycle

2. As stated in the introductory notes, the disaster management cycle can be, and often is, portrayed in various forms. Moreover, alternative terminology may be used. The important factor, however, is that the format should indicate that disaster and managing it is a continuum of interlinked activity. It is not a series of events which start and stop with each disaster occurrence.

### 3. *Basic Format*

The basic format of the disaster management cycle that is used in this handbook, and which is also shown in the introductory notes, is repeated here for ease of reference (figure 1).

### 4. *Alternative Format*

An alternative format, which is sometimes used, shows the main components in the form of activity segments, as shown in figure 2.

It will be noted that the outer segments of preparedness, response, and recovery, superimposed on figure 2, relate to similar segments in figure 1.

**Figure 1: Basic Format of the Disaster Management Cycle**

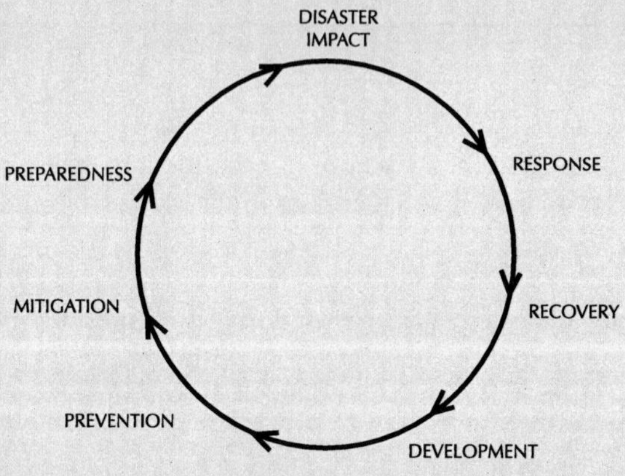


**Figure 2: Alternative Format of the Disaster Management Cycle**

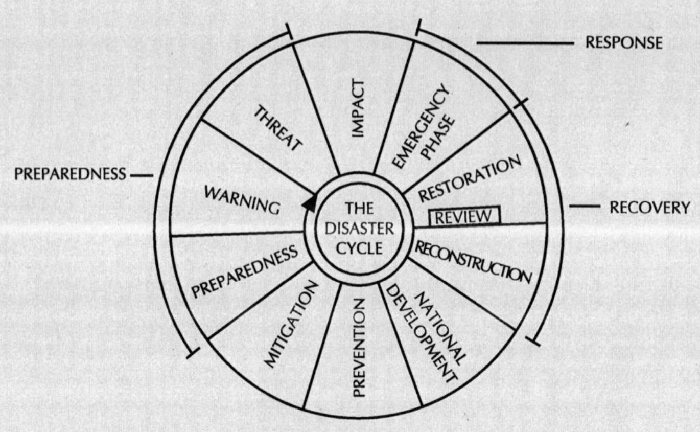

The format in figure 2 can be utilized to make two significant points:

i. Such format is schematic only. It does not and cannot designate the length or relative importance of the component parts. For instance, the actual recovery period may vary considerably for different disasters. Or, in a particular set of circumstances, the amount of importance, priority, and effort allotted to prevention may be small when compared with that given to, say, preparedness.

ii. Such format should not be allowed to give the impression that each activity segment is clearly and precisely divided from adjacent ones. On the contrary, it is important to understand that segments generally tend to overlap and/or merge. For instance, some response activities may be initiated prior to disaster impact—that is, during the preparedness segment. Such activities might include the precautionary movement of threatened persons or communities to safe havens prior to the impact of a cyclone. Similarly, recovery action often begins while the emergency response period is still operative. For example, a technical advisory team would probably begin collecting information immediately after impact and such information would be used for both response and recovery purposes.

Points such as these are important when it comes to practical disaster management action. This aspect is referred to again in paragraphs 13 and 14.

## Composition of Main Activities

5. The composition of the main segments within the disaster management cycle is described briefly in paragraphs 6–12. In this connection, it is worth bearing in mind that two major factors are likely to trigger action in some or all of these segments. These factors may also affect the balance between activities and the priorities allotted to individual activities. The factors are:

- *Post-disasterr eview*
  Post-disaster review should be carried out as early as practicable in the recovery period. Such review will often reveal deficiencies in plans and will also indicate, for example, if certain activities such as preparedness measures and response arrangements need strengthening.

- *Results of exercise or simulations*
  Provided exercises and simulations are accurately evaluated and the lessons from them are correctly drawn, they can exert influences

similar to those of post-disaster review. In some cases, exercises and simulations can be more effective because

- they can be directed toward testing a particular part within the disaster management cycle (e.g., coordination in the use of resources); and
- their lessons can be more accurately defined than is sometimes the case with post-disaster review (because the latter may lack vital information which was overlooked or lost under the pressures of disaster impact).

Furthermore, effective day-to-day disaster management (for example, under an active NDC and NDMO) should monitor all aspects of activity and initiate necessary action accordingly.

6. ***Prevention*** (Chapter 14)

Action within this segment is designed to impede the occurrence of a disaster and/or prevent such an occurrence having harmful effects on communities or key installations. The following are usually classified as preventive measures:

- constructing a dam or levee to control floodwaters so that the latter cannot adversely affect people, buildings and other installations, livestock, means of production and subsistence, and so on;
- controlled burning off in a bushfire-prone area prior to the high fire-risk season. This action can remove potential fuel and actually prevent the start of a fire or, if it does start, prevent it from reaching threatening proportions; and
- some forms of legislation can also be regarded as prevention (e.g., land-use regulations which ensure that communities are not allowed to develop on vulnerable sites such as the disaster-prone areas of a flood plain).

It is noteworthy that some countries tend to use the term prevention/mitigation as a combined heading for action within these two segments.

7. ***Mitigation*** (Chapter 15)

Action within this segment usually takes the form of specific programs intended to reduce the effects of disaster on a nation or community. For instance, some countries regard the development and application of building codes that can reduce damage and loss in the event of earthquakes and cyclones, as being in the category of mitigation. Other countries may regard such building codes as being in the category of prevention; recent developments in earthquake-proof buildings have undoubtedly influenced this outlook.

The term mitigation more generally implies that while it may be possible to prevent some disaster effects, other effects will persist but can be modified or reduced provided appropriate action is taken.

The foregoing points suggest that, under some circumstances, the term prevention/mitigation may be more suitable for some countries than using prevention and mitigation as two separate concepts and activities.

The following actions or programs are generally regarded as coming under the heading of mitigation:

- enforcement of building codes;
- land-use regulations;
- safety regulations relating to high-rise building, control of hazardous substances, etc;
- safety codes governing land, sea, and air transport systems;
- agricultural programs aimed at reducing the effects of hazards on crops;
- systems to protect key installations such as power supplies and vital communications; and
- developments in infrastructure, such as the routing of new highways away from disaster-prone areas.

## 8. *Preparedness* (Chapter 16)

Preparedness is usually regarded as comprising measures which enable governments, organizations, communities, and individuals to respond rapidly and effectively to disaster situations. Examples of preparedness measures are:

- formulating and maintaining valid and updated counter-disaster plans which can be brought into effect whenever required;
- special provisions for emergency action, such as evacuating populations or moving them temporarily to safe havens;
- providing warning systems;
- emergency communications;
- public education and awareness; and
- training programs, including exercise and tests.

One aspect of preparedness that is not always prioritized adequately is individual and/or family preparedness. In many circumstances where government resources and emergency services are limited, such individual and family preparedness may be vital for survival.

Some disaster management cycles may divide the preparedness segment into subsegments such as:

- *Warning* – The time or period when a hazard has been identified but is not yet threatening a particular area (e.g., notification that a cyclone exists but is far away).

- *Threat* – The time or period when a hazard has been identified and is assessed as threatening a particular area (e.g., a cyclone is tracking toward that area).

- *Precaution* – Action taken after receipt of warning to offset effects of disaster impact. Such action might include:
  - closing offices, schools, etc.;
  - bringing emergency power generators to readiness;
  - cutting crops to avoid total loss from high winds and heavy rain;
  - making safe boats and vehicles; and
  - taking household precautions, such as storing emergency water supplies.

An advantage in including these subsegments is that it provides some indication of the possible sequence of events/action leading up to disaster impact.

## 9. *Disaster Impact*

This segment is self-explanatory, being the point in the disaster cycle at which a disaster occurs; for instance, when a cyclone strikes a country or a particular area. However, including it serves as a reminder that—in disaster management terms—impact can vary between different types of disaster. For instance:

- An earthquake may give no warning and its impact time can be short. Yet the result can be very severe indeed.
- A cyclone may provide a long warning period and its impact time (i.e., the time when it has destructive and damaging effects) can be protracted. This may be particularly so if the cyclone passes directly over a given area, or backtracks as may sometimes occur.

10. ***Response*** (Chapter 17)

Response measures are usually those which are taken immediately prior to and following disaster impact. However, for ease of representation, the response segment is shown (figure 2) as following directly after disaster impact; and this is when most response measures are applied.

Such measures are mainly directed toward saving life and protecting property, and to dealing with the immediate disruption, damage, and other effects caused by the disaster. Typical measures include:

- implementing plans;
- activating the counter-disaster system;
- search and rescue;
- providing emergency food, shelter, medical assistance, etc.;
- surveying and assessing; and
- evacuating.

The segment is sometimes called emergency response to indicate that it applies to a fairly short period (i.e., the 2–3 weeks after impact) when emergency measures are necessary to deal with the immediate effects of a disaster and when, perhaps, a state of emergency or state of disaster may have been declared by government.

It may be noteworthy here that it is sometimes said that all disaster-related activities that follow impact (including measures of relief, rehabilitation, restoration, and reconstruction) constitutes response. However, for a user handbook, it is more convenient and practicable to divide response from recovery.

11. ***Recovery*** (Chapter 19)

Recovery is the process by which communities and the nation are assisted in returning to their proper level of functioning following a disaster. The recovery process can be very protracted, taking 5–10 years, or even more. Three main categories of activity are normally regarded as coming within the recovery segment. These are:

- restoration,
- rehabilitation, and
- reconstruction.

Typical activities include:

- restoring essential services;
- restoring of repairable homes and other buildings/installations;
- providing temporary housing;
- measures to assist the physical and psychological rehabilitation of persons who have suffered from the effects of disaster; and
- long-term measures of reconstruction, including the replacement of buildings and infrastructure that have been destroyed by the disaster.

Post-disaster review should also be included as part of the recovery process. It should take place as soon as practicable after the disaster.

12. ***Development*** (Chapter 6)

The development segment provides the link between disaster-related activities and national development. Its inclusion in the disaster cycle is intended to ensure that the results of disaster are effectively reflected in future policies in the interests of national progress. For instance, to produce the best possible benefits by:

- introducing improved and modernized building systems and programs;
- using international disaster assistance to optimum effect;
- applying disaster experience in future research and development programs; and
- using any other means appropriate to a particular situation.

At the same time, this linkage should be used to ensure that national development does not create further disaster problems, or exacerbate existing ones. This point is developed further in Chapter 6.

## Application to Practical Disaster Management

13. It is suggested that individual countries should choose the form of disaster management cycle which is most appropriate to their needs.

14. Apart from its obvious value in providing a "visual aid" for those involved in the study of disaster and in disaster management, the disaster cycle can have various practical applications. For instance, in:

- *Training programs*
  These programs tend to concentrate mainly on the various aspects of preparedness, response, and recovery. Use of the disaster management cycle helps facilitate understanding of not only the important relationship between these three vital aspects but also their connection with other disaster-related activities. The cycle can also have other training applications. For example, it can be used during exercise briefings to illustrate the precise point, within the total disaster management process, at which an exercise is set.

- *Programs of public education and awareness*
  For these programs, the cycle can be used in much the same way as for training programs. It could be especially useful for:
  - disaster education in schools, and
  - heightening public awareness (through posters or television programs) prior to high-risk seasons, such as a cyclone or flood season.

- *Day-to-day disaster management activities*
  The cycle could be an effective reference and calendar against which to check the progress of disaster management at various levels of government; for instance, developing plans, progress in preparedness measures, review by the NDC, and so on.

- *Maintaining government impetus behind disaster management*
  The cycle could be an effective tool during periodic briefings for cabinet and/or those ministers with key responsibilities for disaster management, especially to show where deficiencies need to be remedied.

*Chapter 6*

# Disaster and National Development

## Purpose

1. The purpose of this chapter is to outline the main factors which apply to the interrelationship between disaster and national development.

2. The chapter considers:

- disaster and the national image;
- the impact of disaster on national development; and
- national development and disaster management policy.

## Disaster and the National Image

3. In the modern world, countries are becoming increasingly interrelated and interdependent.

There are many reasons for this, and they include the following:

- world economic framework,
- need to conserve natural resources,
- need to protect the environment, and
- importance of avoiding world conflict with weapons of mass destruction.

Thus, a nation's policies are thrown into sharp relief whenever it becomes involved in major international or national crises. Disaster is one national crisis and it can raise many issues and perceptions that reflect on a nation's image.

## 4. *International Aspects*

When a country is struck by disaster, there is usually widespread international reaction to offer help. This certainly applies when the stricken country is in the developing category and, therefore, has limited capability for response and recovery. It becomes particularly and understandably emphasized when the country is markedly disaster-prone, as with Bangladesh and the Philippines. However, even large and powerful nations may attract international attention and help, as attested to by major earthquake problems in the former Soviet Union in 1988.

There can, however, be damaging consequences from disaster where a nation's international image is concerned. This can apply even despite humanitarian sympathy and willingness to help. Some salient points, based on case studies, are given below.

- International donor countries and organizations have, over many years, provided help to developing countries to help the latter cope with disaster. This assistance has included:

  - regional seminars and workshops aimed at enhancing disaster management knowledge and encouraging the development of counter-disaster capability;
  - providing funding and expertise for the formulation of plans;
  - providing key systems and facilities such as emergency communications, emergency operations centers, and warning networks; and
  - providing training for key officials and others.

  In some cases, recipient countries have made good use of this international assistance and have strengthened their counter-disaster capabilities accordingly. Other recipient countries have not been able to avail of such assistance successfully. In these latter cases, there may have been valid reasons why international assistance could not be fully used. However, this is often lost from sight and the nations concerned may be dubbed as incompetent and not deserving of further assistance.

- In response operations, similar circumstances may arise. A stricken country which is well prepared for disaster usually has a good idea of what its immediate post-impact assistance needs will be. This especially applies to key commodities such as food, clothing, shelter materials, medical assistance, and emergency equipment.

Requests to international assistance sources are, therefore, likely to be reasonably accurate and economical. The task of funding agencies is simplified accordingly.

However, in the case of an ill-prepared nation, difficulties are likely to arise in defining what type and scale of international assistance is required. Misunderstandings and irritations may then arise between the recipient nation and the funding agencies, and these may well have a subsequent adverse effect on the nation's international standing and image.

- It is notoriously difficult to define the exact shape and size of post-disaster recovery programs. This is due to many factors, including:

  - inadequate information from survey and assessment;
  - pressures of post-disaster problems, such as urgent rehousing needs;
  - difficulties in deciding what priorities have to be given to the various individual recovery programs;
  - sensitivities in international dealings with development partners; and
  - post-disaster disruption to the government system of the stricken country.

  Often, the stricken country feels pressured into making hasty decisions and commitments, which subsequently lead to misunderstandings and difficulties. The stricken nation may therefore appear in a less than competent light to its helper counterparts.

Factors such as these do not merely tarnish a country's image. They can also have repercussions on national development generally because they may engender a reluctance from the international community to provide help in the case of vital, nondisaster development programs. In other words, if a nation cannot effectively use disaster-related assistance, why should it be expected to optimize other assistance forms?

However, it is not only the less powerful nations whose images may suffer from disaster-related affairs. A large nation can have problems, too. For instance, pollution of waterways, oceans, or the atmosphere that is generated by industrial and other activities of a large nation, may affect lesser neighbors. The Chernobyl nuclear disaster is a case in point.

In brief, therefore, disaster-related events and activities can have a variety of international effects, some of which may adversely reflect on future development.

### 5. *National Aspects*

From an internal viewpoint, disaster can and often does affect a nation's standing, image, and stability. Understandably, a government comes under heavy pressure when disaster strikes. Thus, government is very much in the spotlight where the population is concerned. A bungling performance by government, especially where it affects the basic needs of the people, will usually result in some form of political backlash. A hungry, uncomfortable, and dispirited public is unlikely to be tolerant and forgiving toward incompetent leadership; thus, political leaders have been known to fall when post-disaster reckoning was made.

6. A combination of the international and national effects that arise from disaster can, therefore, be damaging to a nation's prestige and status, and even its future.

## The Impact of Disaster on National Development

### 7. *Effects on Contemporary Development*

It goes almost without saying that disaster can have very serious effects on the contemporary development of a nation. This is especially so in revenue-producing areas and infrastructure. Typical examples include:

- Crops which have been developed over a number of years to produce an export capability may be destroyed or seriously damaged; for instance, the organized growth of coconut and oil palms as a source of copra and palm oil. Such destruction or damage can result in loss of development capital, destruction of production sources (e.g., the trees themselves), loss of processing facilities and equipment, loss of employee housing, and so on.
- Loss of livestock, through an outbreak of animal disease or rural wildfire, can devastate valuable national resources such as meat or wool trades.
- Land inundated by a cyclonic storm surge or tsunami can bring about salinization with severe consequences to both domestic and export food sources. Similarly, drought may cause severe restrictions on wheat growing, with consequent damage to or loss of a valuable export market.
- In the case of infrastructure, losses from disaster can be crippling for ongoing national programs. For instance:

- Damage to harbors and wharves can drastically limit maritime transport capability, thus, restricting export and import activity.
- Loss of aircraft and airport facilities may pose serious constraints similar to those which apply to the maritime aspects mentioned above.
- Loss of roads and bridges may curb important ongoing construction and other programs.
- Loss of buildings and facilities can seriously hamper the conduct of business and commerce, domestically and internationally.

The overall effect of disaster on contemporary progress and development may, therefore, be one which halts a nation in its tracks. This could be compared with the effects of a heart attack, in that it takes some time before the patient can resume business as usual.

## 8. *Effects on Long-Term Development*

The aforementioned contemporary effects can obviously be crippling for ongoing programs. However, some long-term consequences may be equally bad, or even worse. As a simple example, damaged coconut palms may take 7–8 years to rehabilitate. By this time, a valuable export capability may have become outdated, or lost to other countries. Also, some of the wider economic, social and other effects may bear heavily on long-term development.

One major general fact against which long-term effects should be judged is that when disaster strikes, it deprives a nation of many of its current vital resources. These resources have to be renewed or rehabilitated before even moderate development can continue. Some resources are listed below and they come, by way of example, from an ongoing disaster when this chapter was being written (February 1990):

- communications,
- power supplies,
- housing,
- roads,
- water supplies,
- crops and other food sources,
- airport facilities,
- shipping,
- schools, and
- medical and health facilities.

The national effort and cost required to make good this kind of loss in resources must obviously detract from what is available for long-term development. So, the first effect on long-term programs which arises from disaster impact is the economic and material loss, which is likely to cause delays in commencing and/or continuing such programs.

A second major effect may be on the development planning process. Most countries nowadays work to planned periods of development, such as a series of 5-year national development plans. Disaster can seriously upset this sequence of planning and implementation, possibly as follows:

- It may disrupt the current plan.
- It may make it difficult to forecast development progress, and to predict the likely status at the end of the ongoing plan. These factors will then create problems in determining the start-up parameters for the next planning period. (The effect of this aspect will obviously depend on the point within the ongoing plan at which disaster strikes).
- It may raise problems in formulating the scope and content of the next (or follow-on) national development plan because the post-disaster recovery process is difficult to define accurately. In other words, unanticipated delays in recovery may dictate that resources have to be deflected from planned development.

These factors and others that may apply to individual cases, may cause difficulties in the areas of:

- forecasting,
- budgeting,
- planning progress, and
- implementing a program.

Thus, they upset the rhythm of national development and threaten its satisfactory achievement.

A third effect on national development may emanate from limitations which apply to normal international assistance programs. For instance, bilateral assistance programs usually have mutually agreed limitations on scope, size, time, and cost. Thus, a funding nation may agree to undertake a 3-year program of medical and health assistance. If a significant disaster strikes at, say, the halfway stage of such a program, there will be obvious loss factors. In addition to this, however, the funding agency may well—in the interests of good relations—feel compelled to inject emergency assistance to deal with the disaster situation itself. This combination of loss and extra (unanticipated) help means that the overall cost of the program to the funding agency may have escalated considerably. The result is that the funding agency may have no option but to spread the program over a further year, or even more. Enforced adjustments of this nature can often arise. With nondisaster assistance programs generally, it is often common practice that any extra costs incurred by disaster are offset in future segments of such programs. Thus, national development overall can suffer from various forms of restriction.

A fourth effect is the reaction of development partners to the setbacks caused by disaster. The representative of a development partner once said to the writer of this handbook, "The havoc caused by this disaster has been really disappointing for us. After 6 years of running our assistance program, we were just beginning to see productive results. Now, about 70% of our input has literally disappeared. In other words, we have put in 4 years effort for nothing, and now we have to do that 4 years work all over again."

It is not suggested that this kind of setback discourages development partners from continuing their assistance programs. But it does mean that they occasionally suffer heavy losses themselves. This, in turn, may reduce their ability to implement programs, at least in the short term.

Various other factors arising from disaster may also affect long-term development. They include:

- long-term unsuitability of land areas (e.g., as a result of drought or other causes); and
- enforced relocation of communities.

There is also an important underlying national factor; that is, the loss, damage, and disruption caused by disaster can significantly affect a nation's capacity to absorb subsequent assistance programs, at least temporarily. Assistance inputs can then have an unsatisfactory forced-feeding effect which, far from helping development, can actually slow it down.

9. Before leaving the impact of disaster on national development, it is a sad reflection that countries most in need of development assistance tend to suffer most in terms of setbacks. Stronger nations, having more resources and therefore greater resilience, can cope with the impact on development more successfully than their weaker counterparts.

## National Development and Disaster Management Policy

10. It is clear from paragraphs 3–9 that disaster can have far-reaching repercussions on a nation's image. It can also pose serious restrictions on national development. It follows, therefore, that there needs to be a prudent connection between development and disaster management policy. However, this connection does not usually come about automatically. In this regard, there are two important interrelated points.

11. First, most major departments within government have a tangible and practical connection with disaster management. They have clearly defined roles and responsibilities throughout the total process of prevention, mitigation, preparedness, response, and recovery (especially in these last three segments). These roles and responsibilities are formally laid down in the relevant counter-disaster plans.

12. Second, those government departments primarily concerned with development planning do not usually have direct and practical connections with disaster management. These planning departments and sections would certainly be involved in actions affecting recovery and—perhaps to some lesser extent—with prevention and mitigation. But, generally speaking, they do not have the active day-to-day association with disaster management which ensures a clear and strong working relationship.

13. These two factors suggest, therefore, that some kind of bridging or linking mechanism is required between national development and disaster management.

14. There are, however, some important prerequisites to an effective linking of disaster management activities with those of national development. In most cases, from the disaster management side, these are likely to include:

- The need for the government to recognize clearly, at the highest level, the repercussions which disaster can have on the nation generally and on national development in particular. (If this recognition does not exist, or is not readily forthcoming, then it is obviously up to those who have the responsibility for dealing with disaster to promote such recognition by whatever means are feasible. This needs to be done realistically, not in an exaggerated form; otherwise, it will not achieve credibility).
- The need for a clear definition of disaster management policy,

- Clearly allocating disaster management responsibility at ministerial level.
- Establishing and maintaining an effective disaster management capability.
- A national disaster management office or section capable of maintaining a sound day-to-day liaison with other government departments and agencies.

Obviously, it is assumed that similar prerequisites must exist from the national development side. Also, that a meaningful working relationship applies throughout the ministerial and departmental processes of government.

## 15. *Possible Linking Arrangements*

In most countries, the ministerial portfolios involving national development and disaster management are those covering:

- national development,
- the economy, ] sometimes combined
- finance,
- the environment, and
- disaster management (which, in practice, may be included in portfolios such as police and emergency services, home affairs, community services, defense force, etc.)

Given that the appropriate portfolios are identified and earmarked, the choice of a linking system or arrangement becomes one of national option. One choice considered by a small nation is shown as option A.

In option A, action at provincial government level is processed through the governor and his staff, as for other affairs of government.

A second option, shown as option B, is schematic, based on the structure used in a developing country which is geographically much larger than the example used in option A.

16. The organizational options shown are, of course, for illustrative purposes only. In individual countries, many different factors bear on the relationship between national development and disaster management policy. These factors need to be taken into account when organizational arrangements are made.

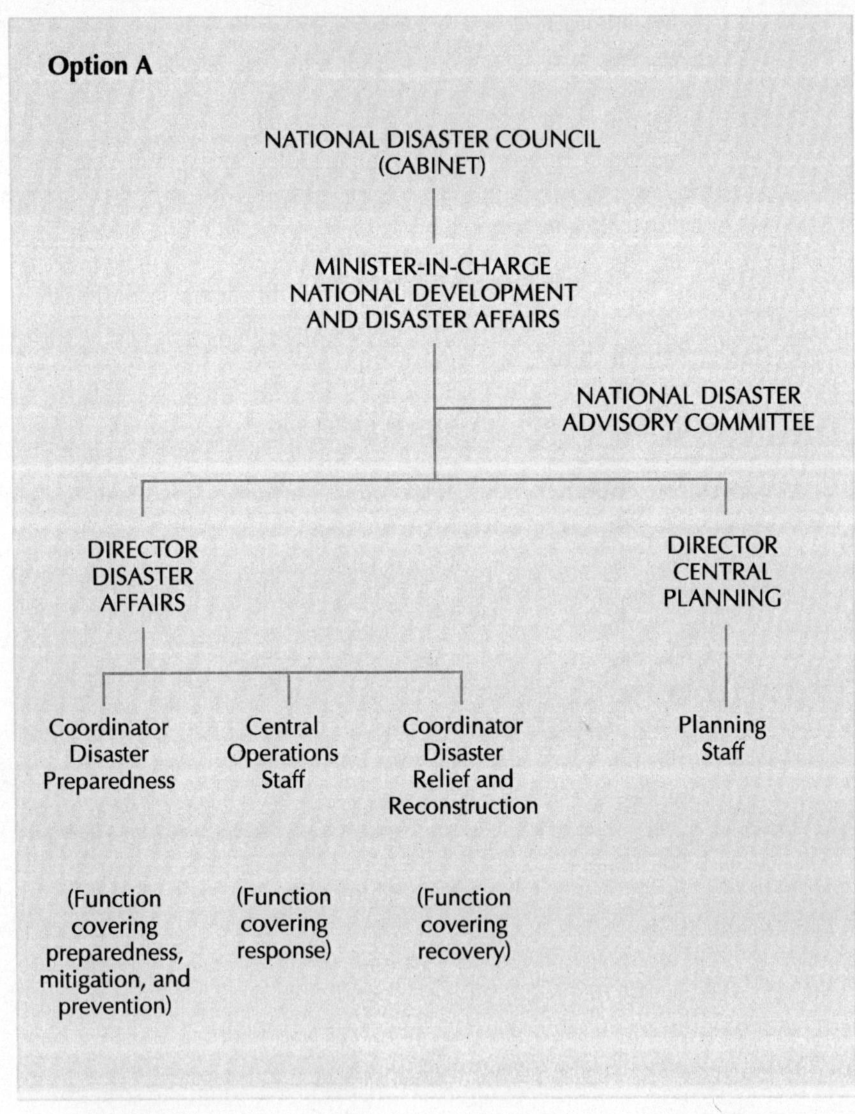
Option A
NATIONAL DISASTER COUNCIL
(CABINET)
MINISTER-IN-CHARGE
NATIONAL DEVELOPMENT
AND DISASTER AFFAIRS
NATIONAL DISASTER
ADVISORY COMMITTEE
DIRECTOR
DISASTER
AFFAIRS
DIRECTOR
CENTRAL
PLANNING
Coordinator
Disaster
Preparedness
Central
Operations
Staff
Coordinator
Disaster
Relief and
Reconstruction
Planning
Staff
(Function
covering
preparedness,
mitigation, and
prevention)
(Function
covering
response)
(Function
covering
recovery)

## Option B

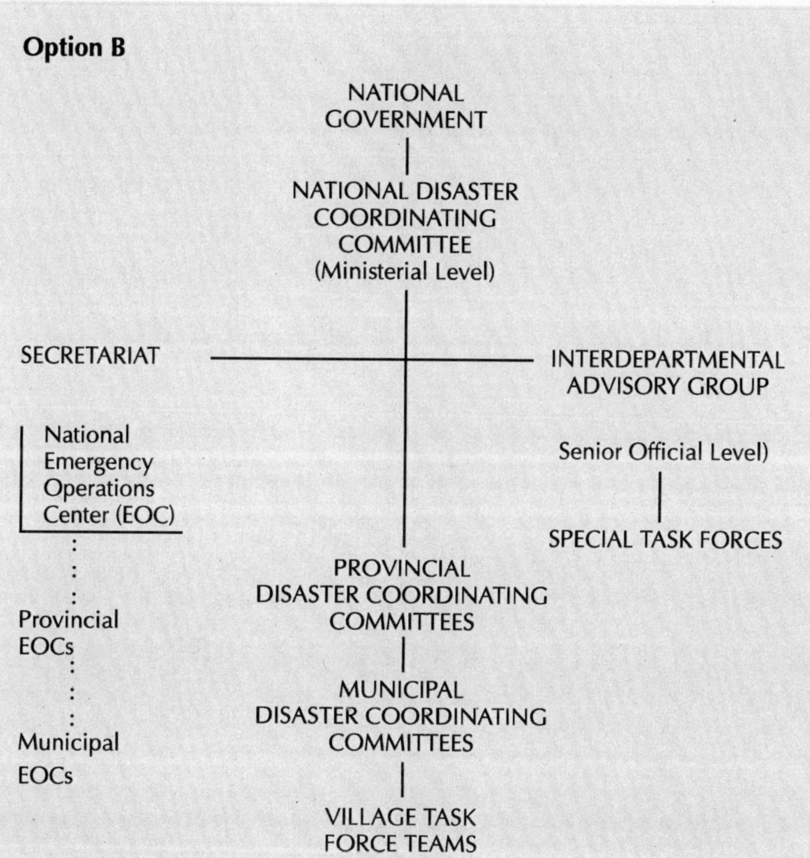


Explanatory Notes on Option B

1. This disaster management structure is directly related to and functions in line with the ongoing 5-year national development plan.

2. The national disaster coordinating committee consists of senior ministers, all of whom have direct disaster management responsibilities. The committee is empowered to coopt other ministers, as required, to deal with special issues.

3. The interdepartmental advisory group, consisting of senior government officials, is required to monitor and advise on all disaster-related matters, including those which affect the relationship between disaster and national development. The goup is empowered to set up special task forces, as required.

4. Provincial disaster coordinating committees and municipal disaster coordinating committees have membership which is functionally in line with the national disaster coordinating committee.

5. An important factor is that the governor of the province is chairperson of the provincial disaster coordinating committee and he has a wide range of responsibilities at his level, including development.

*Chapter 7*

# Disaster Legislation

## Purpose

1. The purpose of this chapter is to outline major factors which apply to disaster legislation. The chapter covers the following:

- need for and value of legislation;
- main areas covered by legislation, using a first draft as an illustration; and
- examples of existing legislation.

## Acknowledgment for Use of Existing Legislation

2. The chapter quotes existing legislation of the Cook Islands, Papua New Guinea, and the State of Queensland in Australia.

3. Thanks are expressed to these governments for permitting their legislation to be included in this handbook.

## The Need for and Value of Legislation

4. Chapters 1–6 of this handbook cover only some limited aspects of disaster management. However, they are sufficient to illustrate the wide repercussions which disaster can have, both nationally and internationally. For instance, a random sample taken from the aforementioned chapters indicates that disaster may cause:

- the fall of governments because of inadequacies in disaster management;
- national loss and damage on a crippling scale;
- problems with international assistance programs;
- delays in national development;
- severe human trauma, hardship, and suffering; and
- widespread destruction of and damage to the environment.

5. A large proportion of disaster-related matters are outside the normal pattern of life. While many, such can be predicted with reasonable accuracy (e.g., the effects of a particular type of disaster), others may be unforeseen. To deal effectively with disaster, therefore, requires a carefully calculated and accurate approach, culminating in a series of countermeasures which involve the government, nongovernment agencies, private sector, and the general public. Seen in this combined and concerted light, counter-disaster action would certainly seem to benefit from an appropriate form of legal backing. In fact, there are very many instances on record where the lack of disaster legislation caused problems and difficulties. Conversely, no cases are known where the existence of such legislation proved to be a hindrance or constraint.

6. Some of the values of legislation are as follows:

- Legislation provides a formal basis for counter-disaster action generally. In other words, it formally supports plans, organizational arrangements, preparedness measures, response action, and so on.
- Legislation allocates major responsibilities in legal form. This helps ensure that such responsibilities will be properly implemented.
- Legislation can be made to have a uniform national effect, thus ensuring that all levels of the national counter-disaster structure receive the full benefit of its support.
- Disaster legislation does not need to be complicated to be effective. It provides commonsense backing for commonsense requirements.
- Legislation provides a wide measure of protection for:
  - governments that are usually charged with the responsibility for safeguarding the nation and its citizens, as far as possible, from the effects of disaster;
  - organizations and individuals who are required to carry out counter-disaster activities; and
  - organizations and individuals who may be affected in various ways by disaster.
- Disaster legislation does not have to conflict with other legislation, such as emergency powers.

## Main Areas Covered by Legislation

7. Disaster legislation can vary according to differing requirements and national systems. However, certain main areas are common to most circumstances, as will be evident from the factual examples quoted later in this chapter.

8. At the same time, a common factor in producing legislation is some initial difficulty in deciding exactly what is to be covered. A useful approach, therefore, is for disaster management officials to define this coverage in the first instance and then for legal officials to convert it into legal form.

9. Below is an example of the initial draft of a national disaster act, as produced by disaster management officials. This is, of course, only one way of presenting such a draft. Its purpose is to provide a general guide only.

## Sample Draft of a National Disaster Act

**DRAFT**

**First Revision**
**10 Nov 1998**

**GOVERNMENT OF EXLAND**
**NATIONAL DISASTER ACT 1999**

**PART 1 – PRELIMINARY**

**Purpose**

1. The purpose of this Act is to provide for the organization and management which is necessary to ensure mitigation of, preparedness for, response to, and recovery from disasters.

**Commencement**

2. This Act comes into force when approved by Parliament.

**Definitions**

3. In this Act, the following definitions apply:

"Agency" means a government agency or a nongovernment agency.
"Disaster" means the actual or imminent occurrence of an event which endangers or threatens to endanger the safety or health of any communities or persons in Exland. or which destroys, or damages, or threatens to destroy or damage any property in Exland, arising from:

(i) cyclone;
(ii) flood;
(iii) tsunami;
(iv) earthquake;
(v) volcanic eruption;
(vi) drought;
(vii) air disaster;
(viii) maritime disaster;
(ix) major civil accident (such as a major fire, bushfire, or explosion);
(x) plague or epidemic; and
(xi) any other similar natural or man-made event, except those detailed in paragraphs 4 and 5.

"Government agency" means:

(i) any body constituted by or under any Act for a public purpose;
(ii) any member or officer of such a body; and
(iii) any person in the service of the Government of Exland upon whom any function, power, duty, or responsibility is conferred by or under any Act.

"Minister responsible" means the minister appointed by the council of ministers to be responsible to that council for all disaster-related matters.

"National disaster plan" means the plan formulated and revised from time to time, as necessary, which defines the action to be taken to deal with disasters in Exland.

Etc.

[*Handbook Note*: This list of definitions needs to be carefully defined from a disaster management viewpoint so that all possibility of doubt, misunderstanding, or misinterpretation is avoided.]

**Limitations**

4. This Act does not authorize the taking of measures to control civil disorders or to bring to an end an industrial dispute.

5. This Act does not cover circumstances arising from combat against an enemy, unless specifically authorized by the council of ministers.

[*Handbook Note*: Governments are usually careful about separating disaster circumstances from such matters as breaking of strike and war operations.]

## PART 2 – ORGANIZATION

### Minister Responsible

6. The council of ministers shall appoint a minister responsible to the council for all disaster-related matters applicable to this Act. In particular, the minister responsible is:

(i) to ensure that adequate measures are taken by government agencies to mitigate, prepare for, and respond to disasters and to assist in the recovery from the effects of a disaster by persons or communities;
(ii) to coordinate the activities of government agencies carrying out their statutory functions, powers, duties, and responsibilities in taking such measures; and
(iii) to foster and facilitate the participation of nongovernment agencies in measures taken by the government for disaster mitigation, preparedness, response, and recovery.

### National Disaster Council

7. There is hereby established under this Act a body to be called the national disaster council to advise the minister responsible on all matters, including the coordination of activities of government and nongovernment agencies, relating to disaster mitigation, preparedness, response, and recovery.

8. The council will consist of:

(i) a chairperson nominated by the minister responsible from within the minister's department;

(ii) a representative from each agency which the minister responsible considers should be so represented, to be nominated:

(a) in the case of a government agency, by the appropriate minister: and
(b) in the case of a nongovernment agency, by the agency.

[*Handbook Note* Some countries prefer to nominate the holders of specific appointments as members of the national disaster council. However, the method shown above allows for considerable flexibility. Therefore, if changes to the council become necessary, there is no need to alter the Act.]

9. The national disaster council is to establish a central operations group to assist it during emergency operations carried out in response to a disaster. The role of this group is to ensure, on behalf of the council, that resources are allocated to operational tasks in the most effective way and in the correct priorities.

10. The procedures of the national disaster council are to be as determined by the chairperson.

**National Disaster Management Office**

11. There is also established under this Act a national disaster management office to carry out disaster management and other responsibilities as determined by the minister responsible and/or the national disaster council to advise the minister and the council, and to deal with routine disaster-related affairs at national level.

**Local Government Disaster Committees**

12. Under this Act, each local government council is required to establish a disaster committee.

**Area Disaster Committees**

13. Each area council is also required to establish a disaster committee. These area disaster committees are responsible for coordinating disaster-related measures in their own areas with those of local government disaster committees and the national disaster council.

**Committees Established by the Minister Responsible**

14. In addition to the organizational arrangements outlined in paragraphs 7–13, the minister responsible may establish such

additional committees as are necessary to ensure comprehensive and integrated disaster management.

[*Handbook Note:* The amount of organizational detail to be included in legislation is a matter of individual country choice. In the example above, only the bare outline has been included on the assumption that additional details can be given in the national disaster plan and associated plans.]

## PART 3 – PLANS AND PROCEDURES

### National Disaster Plan

15. There shall be a national disaster plan, approved by the council of ministers. The plan is to define the action to be taken to deal with disasters in Exland, covering all national land and sea areas. The plan is to cover requirements for disaster mitigation, preparedness, response, and recovery.

16. The chairperson of the national disaster council is to be responsible for ensuring that the plan is periodically reviewed and updated as necessary.

17. Government agencies and nongovernment agencies which are formally allocated roles under this plan are responsible for making their own plans and other arrangements necessary to fulfill such roles.

18. Also under this plan, all agencies involved are required to report annually, at a time specified in the plan, to the national disaster council, notifying their state of readiness for the coming year.

### Local Government and Area Councils

19. Local government and area councils are to issue such instructions as may be considered necessary to support the provisions and requirements of the national disaster plan.

## PART 4 – DECLARATION OF STATE OF EMERGENCY

20. If the circumstances of a disaster are, or appear likely to become beyond the scope, provisions, and resources of the national disaster plan, or for any other reasons associated with a disaster, it may become necessary to declare a state of emergency. Provision for

the declaration of such a state of emergency is contained in Chapter 13 of the Constitution. A state of emergency may apply to part or the whole of Exland.

[*Handbook Note*: Most national constitutions contain provision for the declaration of a state of emergency. However, where this is not so, such provision can be included in disaster legislation.]

## PART 5 – SPECIAL POWERS DURING A STATE OF EMERGENCY

21. If a state of emergency is declared in accordance with Part 4 of this Act, the minister responsible—normally on the advice of—will recommend to the council of ministers action which may be necessary, including the making of regulations, concerning some or all of the following aspects:

- (i) supply and distribution of food, water, fuel, power, medical assistance, shelter, and other necessities;
- (ii) maintenance of the means of transportation by land, air, or water and the control of the transport of persons and things;
- (iii) temporary acquisition or control of any property, undertaking, or land;
- (iv) evacuation of people in their own interest and/or that of the public;
- (v) entering and search of any premises; and
- (vi) payment of compensation and remuneration to persons affected by government action under special powers; and
- (vii) any other matters which may apply to a specific disaster event.

## PART 6 – MISCELLANEOUS

[*Handbook Note*: As its name indicates, this section of disaster legislation is used to bring together a number of the smaller, but nonetheless important, disaster-related matters which benefit from legal backing. The examples given below are fairly typical of items included under this heading, but see existing examples of legislation in appendix D for other items and methods of presentation.]

### Offense of Obstructing a Disaster Worker

22. A person must not obstruct, hinder, or in any way interfere with a person engaging in any activity as a member officer or volunteer of an agency performing a role or discharging a responsibility in accordance with the national disaster plan, or in accordance with any regulations made under a state of emergency.

### Immunity from Liability

23. A person engaging in any activity as a member officer or authorized volunteer worker of an agency performing a role or discharging a responsibility in accordance with the national disaster plan shall not be liable in respect of any loss or injury sustained by any other person, unless such loss or injury is caused by or arises from negligence or willful default.

### Claims for Compensation and Remuneration

24. Any person who claims compensation or remuneration for any action taken under the national disaster plan and/or any associated regulations has the right to submit such a claim or claims to the Government of Exland.

### Offense of Making a False Compensation Claim

25. A person must not make any false or misleading statement, or otherwise attempt to mislead, in making any claim for compensation in relation to the provisions of this Act.

### Use of Disaster Assistance Donations

26. All donations, in money or any other form, which are made either from within Exland or overseas, to help in recovery from disaster, or for other disaster-related needs, shall be used for these purposes only. Such donations shall not be diverted for any other purpose without the authority of the council of ministers and the specific agreement of the donor persons, agency, or country.

### Regulations

27. The council of ministers may make regulations consistent with this Act for or with respect to all matters required or permitted by this Act and all matters which, in the opinion of the council of ministers,

are necessary or convenient for the proper administration of this Act or to achieve the objects and purposes of this Act.

**Relationship of this Act to Other Acts**

28. Unless specifically provided for under regulations within a declared state of emergency, this Act shall not alter or affect the provisions of other Acts.

10. It is reemphasized that the foregoing draft of a National Disaster Act is meant to be a possible example only.

11. It is suggested that disaster management officials who may be required to advise on legislative requirements should use the above example in conjunction with a study of the legislation of the Cook Islands, Papua New Guinea, and Queensland which are contained in appendix D (and/or with any other examples of suitable legislation).

*Chapter 8*

# Counter-Disaster Resources

---

## Purpose

1. The purpose of this chapter is to consider the resources that are available in most countries for disaster management purposes.

2. As stated in Chapters 3 and 4, the effective use of available resources is a major disaster management objective. To achieve this objective, it is evident that there must be:

- accurate identification of resource;
- correct assessment of their capability;
- allocation of suitable roles to resource organizations; and
- plans and procedures to use resource in a timely and effective manner.

3. The chapter therefore covers:

- types of resources, both national and international;
- evaluation of resource; and
- roles and responsibility of resource.

4. The information in the chapter has been drawn from a number of different countries and is intended to provide a general checklist. Obviously, however, users of this handbook need to identify and assess their own resources, and allocate roles and responsibilities.

## Types of Resources

### 5. *National resources*

*Government resources*

Departments, section, and agencies covering the following responsibilities:

- Ministry Responsible for Disaster Affairs,
- Police,
- Agriculture,
- Ambulance,
- Attorney-General,
- Audit,
- Broadcasting,
- Civil Aviation,
- Customs and Excise,
- Education,
- Electricity Commision,
- Environment,
- Firefighting Service,
- Finance,
- Fisheries,
- Foreign Affairs,
- Forestry,
- Immigration,
- Information,
- Labor,
- Lands and Survey,
- Marine,
- Medical and Health,
- Meteorology,
- National Development Planning,
- Natural Resources,
- Post and Telecommunications,
- Prime Minister's Office,
- Public Service Commission,
- Public Works,
- Statistics,
- Social Welfare,
- Transport, and
- Treasury.

*Nongovernmentr esources*

- Airlines,
- Amateur radio operators,
- Aviation resource (flying clubs, private owners, etc.),
- Chambers of Commerce,
- Church and religious groups,
- Community service clubs,
- Food supplier (wholesale and retail),
- General agents,
- public (volunteers, blood donors, etc.),
- Red Cross/Red Crescent,
- Transport companies (private: land and sea), and
- Welfare organizations.

## 6. *International Resource*

(Chapter 9 on international disaster assistance)

The use of international resources by any country is likely to depend on factors, such as:

- scale and effectiveness of its own resources;
- bilateral assistance arrangements with allied countries;
- political implications, from a recipient country's viewpoint of accepting international assistance; and
- suitability of international resources in a particular disaster situation.

The type of international resources most generally needed by stricken countries fall into the following broad categories:

- *Pre-disaster*
  Equipment and expertise for developing measures of prevention, mitigation, and preparedness; for instance, providing emergency communications or developing plans and organizational arrangements.

- *During response operations*
  - Monitoring and warning of a potential threat; and
  - Post-impact survey and assessment; for instance, by photographic reconnaissance from the air. Providing emergency equipment and facilities such as shelter materials, food supplies, and medical assistance and, if appropriate, personnel to help distribute and use this equipment and facilities.

- *During recovery programs*
  Technical and other assistance, usually in the form of building materials, financial grants, agriculture redevelopment programs, etc.

- *For future development*
  Assistance from the resources of overseas governments and international organizations, often as an extension of recovery programs.

Source of international assistance may vary for individual countries but usually comprise:

- Multilateral organizations and agencies, and
- Bilateral organizations

## Evaluation of Resources

7. It is important to evaluate national resource organizations to determine their suitability and effectiveness for disaster management purposes. This can be done under the following headings:

### *Capability*

The ability of resource organizations to carry out designated disaster-related tasks; for instance, first aid, search and rescue, emergency feeding, building repairs, welfare assistance, counselling of victims and disaster workers, and so on. In this regard, the following need to be considered, such as:

- Resource organizations produce their best results when used in their most applicable and experienced roles. It is obviously not prudent to make fundamental changes to the role of a resource organization when it switches to a disaster role.
- Some resource organizations are, by their nature, crisis-oriented; for instance, police, firefighting services, ambulance services, and defense forces. Other resource organizations are not normally crisis-oriented such as welfare departments. Yet these departments may have to work under very difficult and traumatic conditions in providing help to disaster victims. Their capability may, therefore, be reduced during disaster operations, especially if their members are inexperienced. This has happened on various occasions.
- The disaster role capability of resource organizations may vary from time to time. This may arise from changes at leadership levels and/

or extensive turnover of staff, both of which tend to lower overall experience within the organization.

- The standard role of a resource organization may change for various reasons, such as government policy and financial constraints. This may affect its capability in its disaster role.

The above factors indicate not only the need for careful basic evaluation of capability but also the need to monitor and reevaluate as circumstances change.

### *Availability*

This is primarily a question of whether the resource organization is immediately available to undertake its disaster task, or whether there is a lead time (definite or indefinite) before it can respond.

A number of different factors may affect availability such as:

- nondisaster (or normal) role of organizations. For instance, a hospital is unlikely to be able to respond with all its resources to a disaster impact. It would have to continue normal responsibilities— even if reduced and—therefore, its disaster availability might have to be made on a graduated scale. This consideration indicates that, in some cases, a gradual availability of resource organizations may have to be recognized and accepted;
- availability of volunteer organizations may be subject to some initial delays because of difficulties in mobilization; and
- internal arrangements within resource organizations for switching over to disaster roles will also affect availability time scales.

### *Durability*

This concerns the degree to which resource organizations can carry out sustained operations and when they need to be relieved by other organizations. Some considerations which apply to this aspect are:

- operational circumstances such as extreme weather conditions, limited resources, heavy task loads, high levels of damage and disruption, and community trauma and suffering.
- type of resource organization. A crisis-oriented organization is likely to be highly trained and well equipped. It is usually independent (i.e., with own communications, transport, feeding arrangements, and so on). Also, it is likely to have its own system for rotating

personnel. Such crisis-oriented organizations are likely to have high-durability levels. Organizations which are not crisis-oriented are likely to have lower durability; and
- type of operations involved. Some operations (for example, search and rescue in a major building collapse or train collision) can be very demanding physically and traumatic mentally. Others are less so.

Durability assessment of resource organizations obviously needs to take these factors into account.

### *Operational integrity*

This concerns the ability of a resource organization to undertake an allotted task and complete it satisfactorily without supervision or continuous detailed direction from the disaster direction/coordination authority. This is a significant factor in the effectiveness of a resource organization, since it can be relied on to:

- accept a task,
- complete it, and
- on reporting completion, be ready to accept retasking.

### 8. *The Process of Evaluation*

The process of evaluating resource organizations should be carried out mutually between the disaster management authority (e.g., NDC or its NDMO) and the resource organization itself. Experience has shown that if the disaster management authority tries to make a general evaluation without consulting the resource organization, the evaluation will be inaccurate and misleading.

## Roles and Responsibilities of Resource Organizations

9. As stated earlier, the information contained in this chapter has been drawn from patterns used in a number of different countries. The roles and responsibilities detailed in paragraph 10 below will therefore need to be adjusted to the circumstances of individual countries. However, they do indicate the wide extent to which both government and nongovernment agencies should become involved in disaster management.

10. In the following examples, it is assumed that there is an NDC and a Central Control Group (CCG). Some suggested allocation of roles and responsibilities are:

## Government Departments, Sections, and Agencies

*Ministry Responsible for Disaster Affairs*

- Responsible for overall planning, organization, training, public awareness, and administration in relation to disaster affairs;
- Provides a permanent secretary as chair of NDC;
- Administers NDMO;
- Maintains liaison with provincial authorities on day-to-day disaster management matters; and
- Provides administrative services necessary for functioning of NDC.

*The Permanent Secretary*

- Responsible to minister for all disaster-related matters;
- Acts as chair of NDC; and
- Directs activities of NDMO.

*Police*

- Commissioner of police acts as member of NDC and chair of CCG;
- Assistant commissioner of police (Operations) acts as deputy chair of CCG;
- Responsible, as in normal role, for maintaining law and order and protecting life and property; reinforces critical areas as necessary in times of disaster;
- Uses police mobile force as necessary;
- Undertakes and coordinates survey and assessment duties, as required by CCG;
- Controls movement in disaster areas and at operational key points.
- Undertakes and coordinates land search and rescue;
- Supervises evacuation operations, as required by NDC;
- Develops disaster-related training within the police force; and
- Takes standard police action to deal with dead persons.

*Agriculture*

- Permanent secretary acts as co-opted member of NDC;
- Undertakes survey and assessment of damage to crops, etc.
- Advises on procedures to safeguard crops, livestock, and equipment from effects of disaster;

- Advises and helps disaster victims whose crops and equipment may have been damaged or destroyed;
- Helps in providing replacement seeds and livestock;
- Operates an early warning scheme of food shortage by monitoring crop failures and food prices, to give maximum warning of any forthcoming shortages; and
- Monitors crop programming aimed at minimizing possible damage and destruction during maximum risk seasons.

*Ambulance Service*

- Deploys resources to maximum effect, as advised by CCG, and
- Undertakes first-aid training, as advised by NDC.

*Attorney-General*

- Acts as co-opted member of NDC; and
- Advises on all disaster-related legislative matters.

*Audit*

- Audits disaster-related accounting.

*Broadcasting*

- Director acts as co-opted member of NDC;
- Broadcasts warning and public information, as required by NDC.
- Maintain 24-hour broadcast capability during disaster periods, as notified by NDC/CCG;
- Advises, in broadcast format, of disaster-related warnings and public information;
- Advises on broadcast segment of public awareness programs; and
- Helps, where possible, in promoting public awareness as advised by chair of NDC.

*Civil Aviation*

- Provides controller of civil aviation as member of CCG;
- Maintains plans and deals with aircraft accidents and incidents, in accordance with international air regulations;
- Helps, where possible, with availability of aircraft for disaster operations;
- Makes departmental communications facilities available for disaster purposes, as far as possible.

*Customs and Excise*

- Facilitates entry of all official disaster assistance commodities and waives customs and excise duties;

*Education*

- Permanent secretary acts as co-opted member of NDC;
- On advice of CCG, or on other indications of disaster, takes appropriate action to ensure safety of school children;
- Makes available, if required, school buildings nearest to disaster-affected areas as temporary welfare and evacuation centers (as arranged with NDC/CCG);
- Provides staff as administrative managers of buildings being used as welfare and evacuation centers;
- Coordinates with other agencies in informing people of impending disaster, especially in remote areas;
- Helps, where possible, in assessing damage and reports information to police; and
- Includes disaster awareness aspects in school programs.

*Electricity Commission*

- Manager acts as co-opted member of NDC;
- Ensures, as disaster preparedness measure, that power lines and other installations are kept clear of trees and other possible obstructions to power supplies;
- Maintains power supplies at best possible level during emergency situations;
- Implements public safety measures in danger areas caused by damage to plant installations equipment; and
- Restores disrupted power supplies in accordance with priorities notified by NDC/CCG.

*Environment*

- Permanent Secretary acts as co-opted member of NDC; and
- Advises NDC on environmental issues that may affect disaster management.

*Firefighting Service*

- Advises NDC on fire hazards and fire prevention; and
- Carries out firefighting, rescue, and other appropriate operations.

*Finance*

- Provides permanent secretary as member of NDC;
- Authorizes release of funds to meet immediate disaster needs;
- Reviews—in conjunction with ministry responsible for disaster affairs and ministry responsible for planning—financial requirements for longer-term relief and recovery, including distribution of monies from any national disaster relief fund; and
- Finances stock holdings of disaster-related stores.

*Fisheries*

- Undertakes survey and assessment of damage to fisheries areas;
- Assists in assessment of loss or damage to fishing vessels, facilities, etc;
- Renders advice and assistance on fisheries aspects; and
- Assists in providing boats, where possible.

*Foreign Affairs*

- Provides permanent secretary as member of NDC; and
- Processes, through diplomatic channels, offers of and requests for overseas disaster assistance.

*Forestry*

- Undertakes survey and assessment of damage to forest areas; and
- Renders advice and assistance on forestry matters, such as disposal and use of trees damaged by disaster.

*Immigration*

- Facilitates entry of approved international assistance personnel.

*Information*

- Chief information officer acts as co-opted member of NDC;
- Advises NDC on media liaison and all information aspects; and
- Drafts approved information in form suitable for issue.

*Labor*

- Advises NDC on industrial safety and associated matters.

*Lands and Survey*

- Provides maps and mapping information;
- Provides field teams (or team members) in operations requiring special field knowledge and skills; and
- Identifies areas suitable for resettlement of disaster victims.

*Marine*

- Chief marine officer acts as member of CCG;
- Provides support for sea movement of disaster-related personnel, supplies and equipment;
- Takes action to deal with maritime accidents; and
- Liaises with local shipping agents for support from foreign and local vessels.

*Medical and Health*

- Provides permanent secretary as member of NDC.

- Provides emergency medical treatment of disaster victims and subsequent hospitalization, if necessary;
- Provides medicines and medical supplies;
- Institutes preventive and curative measures to check and control occurrence and spread of disease;
- Exercises supervision of public health;
- Maintains sanitary conditions in disaster-affected area;
- Checks potability of water supplies;
- Coordinates use of medical teams, medical supplies; equipment provided through overseas disaster assistance; and
- Provides medical certification for dead persons.

*Meteorology*

- Director acts as co-opted member of NDC;
- Provides weather-related information of potential disasters and monitors progress, as required;
- Provides information for public awareness programs;
- Liaises with broadcasting service on broadcast of warning and other weather-related public information; and
- Provides advice to NDC on all meteorological matters.

*National Development Planning*

- Permanent secretary acts as co-opted member of NDC;
- Implements where possible, in national programs, measures likely to assist in long-term disaster prevention, mitigation, and preparedness; and
- Takes action, as required by Cabinet, to ensure that recovery programs are compatible with national development policy.

*Natural Resources*

- Permanent secretary acts as co-opted member of NDC;
- Provides information and advice to NDC concerning possibility of earthquakes and volcanic eruptions; and
- Chief geologist acts as co-opted member of NDC.

Posts *and Telecommunications*

- Provides permanent secretary as member of NDC;
- Ensures rapid repair of damaged telecommunications; and
- Provides emergency communications, as notified by NDC/CCG.

*Prime Minister's Office*

- Provides secretary to Cabinet as member of NDC;

*Public Service Commission*

- Permanent secretary acts as co-opted member of NDC; and
- Assists in providing extra personnel during an emergency period, as requested by NDC.

*Public Works*

- Permanent secretary acts as member of NDC;
- Assists with survey and assessment of damage;
- Undertakes clearance of roads, airfields, and port areas;
- Restores inoperative or damaged public installations and facilities;
- Undertakes special tasks as requested by NDC; and
- Develops and implements technical measures, such as physical planning, zoning, and building codes, for mitigating effects of disasters.

*Statistics*

- Collects, collates, and issues disaster-related data and analyses.

*Social Welfare*

- Permanent secretary acts as co-opted member of NDC;
- Has primary responsibility for relief programs (in which it collaborates closely with NGOs' disaster coordinating committee), including the following main aspects:
  - Emergency feeding;
  - Emergency clothing and household supplies;
  - Providing temporary shelter assistance; and
  - Providing and managing government buildings as
  - evacuation and welfare centers.

*Transport*

- Permanent secretary acts as member of NDC;
- Maintains in-country information on availability of transport for use in counter-disaster operations; and
- Coordinates use of transport, as required by NDC/CCG.

*Treasury*

- See Finance

## Nongovernment Organizations (NGOs)

*Red Cross/Red Crescent*

(Working in collaboration with NGOs' disaster coordinating committee)

- Secretary acts as co-opted member of NDC;

- Holds disaster relief stocks and equipment;
- Provides relief items and assistance to disaster victims;
- Undertakes tracing of missing persons;
- Carries out first-aid training and welfare and education programs; and
- Liaises with other NGOs in coordinating relief work.

*Other NGOs*

(Roles and tasks regulated and coordinated by NGOs' disaster coordinating committee)

- First aid;
- Assistance with survey and assessment;
- Clothing and household supplies;
- Help in tracing missing persons;
- Assistance in providing temporary shelter;
- Providing and managing nongovernment buildings as evacuation and welfare centers;
- Assistance in management of government evacuation and welfare centers;
- Assistance with transport;
- Providing temporary storage facilities;
- Assistance with public education and awareness; and
- Assistance with preparedness measures at community level.

11. It is reemphasized that the roles and responsibilities outlined in paragraph 10 are for purposes of illustration and guidance only. They are not necessarily comprehensive or relevant to all government structures. They therefore need to be used selectively in individual situations.

## 12. *Inclusion of Roles in Plans*

Once roles and responsibilities have been established and mutually agreed, they need to be clearly set out in all plans and other relevant documents, such as:

- national disaster plan;
- provincial disaster pan;
- municipal disaster plan;
- local disaster plan;
- departmental operation procedures; and
- standard operating procedures

See also Chapter 12 on plans.

Annex to Chapter 8

# MILITARY FORCES IN THE COUNTER-DISASTER ROLE[1]

## *Purpose of the Annex*

The purpose of the annex is to outline key factors which apply to the role of military forces in counter-disaster activities.

## *Introduction*

2. The contribution which military forces can make in dealing with disaster has long been recognized in many countries throughout the world. This contribution has usually been made under some form of official arrangements for aid to the civil power, as laid down in national constitutions or special legislation.

3. However, experience has shown that if military forces are to be fully effective in this context, their capabilities must be fully understood by relevant counter-disaster authorities; and their roles and responsibilities must be clearly defined in counter-disaster plans

## *Capabilities*

4. The organization and management systems of military forces make them well suited for operation under disaster conditions. In addition, many of their normal activities parallel those in public emergency services. Thus military forces can offer invaluable support in engineering, communications, transport, rescue, emergency medical sevices, field sanitation, cooking, water supply, and so on.

5. The units available to provide these services are usually deployed countrywide and close to centers of population, which are where disaster may have its most significant effect. These military forces have quick reaction capability and can respond rapidly, taking with them their own management, communications, and administrative systems in a self-contained, self-sufficient, and highly mobile form. They are well trained in the individual

1 Much information in this Appendix is drawn from the paper entitled *The Military Role in Disaster Relief, Preparedness and Prevention* by Lieut. Colonel G.N. Ritchie, Director of the Cranfield Disaster Preparedness Centre, United Kingdom and its use in this handbook is gratefully acknowledged.

skills necessary to perform their professional and functional activities and are practiced in collaborative and coordinated action under their own flexible management systems.

6. Add to this the capability to sustain operations away from their home base in all weather conditions by day and by night and one sees clearly the potential inherent in any military organization for effective disaster relief operations.

## *Constraints*

7. Although armed forces do possess the capabilities described above, often there is lack of adequate arrangements in legislation to ensure that they are readily available to support the civil community and administration in response to disasters arising from natural or man-made causes. Matters are usually complicated by the fact that armed forces are outside the normal chain of command and communications which link the system of central and local government and their supporting public and emergency services. The situation may be made more difficult as military commanders may have a different set of responsibilities, and loyalties, from those of the local government officials and administrators requiring their support. Moreover, there is the complication of a different set of financial controls and authorities, budgets and accountancy, procurement, and supply systems.

8. In many countries, military commanders at all levels are unlikely to be practiced in operations with civilian authorities and public services unused to their system of decision making, authority, and responsibility. There is likely to be incompatibility in equipment (radio communications, for example) and little experience of collaboration in achieving a common task.

9. A further factor is the justifiable wariness that civil government in many countries has of military coups and takeover of government. A situation created by a major disaster is one frequently leading to political instability, providing opportunities—possibly even the requirement—for the military to assume authority and responsibility. Examples exist where the military and civilian authorities have been in competition to demonstrate their superior capabilities in responding to disaster situations. Nevertheless it can be argued that the most effective way to avoid such dangers is to integrate the capabilities of a country's military forces with the civil authority's, together with the supporting public services, at all stages and all levels of counter-disaster planning and preparations.

10. This may well require some degree of legislation and financial provision but the benefits to the administration and the people will be much greater

than any bureaucratic inconvenience created. Not only do these military forces make a major contribution to disaster relief operations but the resources of their technical units, particularly engineers, can also make major contributions in disaster prevention.

## *Sources of Military Assistance in Disaster*

11. *Sovereign Military Forces*

There are four principal sources from which military assistance to the civil communities are likely to be available. The first of these, sovereign national forces, is clearly the most readily available and, most importantly, can, as a result of their involvement in pre-disaster planning and preparedness, be most effectively integrated in meeting the responsibilities of the civil administration and its supporting public emergency services.

The regular, full-time military force is probably best trained and equipped for this role, but it is equally probable that it may have other responsibilities and even operational commitments that constrain its availability when its help is most needed. The arguments, therefore, for national militia—that is, part-time military forces—to have a significant role in counter-disaster action are strong. They undoubtedly have the advantage of being a part of the community in which they are located, and have a good and current knowledge of local resources, the operational area, the disaster threat, and its implications. Their officers will know and be known by the officers of the local civil authority as well as public and emergency services. All these will lead to effective degrees of coordination and collaboration arising from the knowledge of the availability of these capabilities, requirements, and constraints affecting the various plans of the availability of these various resources. On the debit side, militia forces may lack the skills and equipment of regular forces and may also lack technical training. Key members of the force may themselves become victims of the disaster or in disaster have family or personal commitments that prevent their availability when most needed. Nevertheless, the role which their organization and training, location, and local commitment can play in disaster is likely to counterbalance these deficiencies.

12. *International Collaboration and Aid*

Currently and historically, there is evidence of the collaboration and coordination that can be achieved in the groupings of military forces in alliances between friendly powers. Although there is much evidence of the effectiveness of such collaboration in military operations, there is also evidence of such collaboration in disaster. For instance, when the "Herald of Free Enterprise" sank in 1988 off the Belgian coast, the levels of collaboration between military units from Belgium and the United Kingdom (UK) with civilian emergency services were a most significant element in the effectiveness of the rescue operation. Similar levels of coordination between naval and air forces of North Atlantic Treaty Organisation (NATO) countries and civilian rescue services were of a very high level during rescue operations after a disastrous oil rig fire in the North Sea.

13. *Military Forces from Friendly Countries*

There have been numerous examples in the past of military assistance—principally in the form of transport aircraft and helicopters—being sent from friendly countries to help disaster-stricken developing countries. In recent years, aircraft of these types from Belgium, Germany, Poland, and the UK have collaborated in air delivery and air dropping operations in Ethiopia and Nepal.

At other times, more particularly in areas such as the Caribbean and the South Pacific, when island communities have been struck by disaster, naval ships have been able to come into a harbor or alongside jetties to provide help in the form of manpower, helicopter reconnaissance and airlift, engineering skills, fresh water, and even baked bread.

Such forms of help are easily negotiated and provided under existing rules and procedures. Foreign aircraft fly into national airfields everyday and foreign ships, including naval ships, come in to take water, fresh food, and other stores. Their presence is contained within the system operating at ports and airports.

However, the introduction of foreign army units that require accommodation—which are often accompanied by large quantities of equipment—that live and work in the community, requires much more careful planning and detailed arrangements. Consequently, experience suggests that, for optimum benefit, army units should be small, self-contained, and self-supporting; also that they remain for a limited time (say, 2–3 weeks during the peak emergency period) in the host country. For example, in the Pacific region, army units of Australia, France, New Zealand, the UK, and the United States (US) frequently provide help on this kind. It is important, of course, that wherever possible such help should be part of mutual disaster-preparedness planning.

## *The Nature of Military Assistance in Disaster*

14. The nature and extent of military assistance in disaster will obviously depend on the combination of in-country and overseas forces that are available. However, the following are typical examples:

- Air, land, and sea survey and assessment, especially to ascertain levels of casualties and damage;
- Reconnaissance in the disaster zone and of the routes within and leading toward it;
- Organized and mobile manpower to help in rescue;
- Well-trained and equipped engineers;
- Communications units to provide emergency radio and telephone links;
- Logistics units available to handle, store, and transport goods and people over all types of roads and routes;
- Emergency medical support to the civilian services;
- Helicopters and aircraft for reconnaissance and transport duties; and
- Provision by naval forces of transport support when in harbor; and electrical power, mechanical engineering, water purification, long distance radio communication, and cooking and baking facilities.

In some disaster-prone countries, military resources have been integrated into the national counter-disaster system, from national to the local community levels, with valuable results. In other countries, however, the absence of joint civil–military planning has prevented the optimum use of military resources.

15. *National Level*

Effective national disaster–preparedness planning should be based upon a logical and realistic analysis of the threat and the vulnerability to the threat of communities, infrastructure, and economic assets. In the consequent disaster-preparedness planning, responsibilities, and tasks in response to disaster occurrence will be identified at various levels of national and local government. As mentioned above, these will include:

- Reconnaissance and reporting upon the damage caused and the state of infrastructure and communications and the use of warning to endangered communities;
- Rescue, evacuation, and providing life support system (medical aid, water, food, and shelter) in emergency centers or camps. Providing safe drinking water and emergency sanitation in such circumstances is of prime importance;
- Road clearance and the repair of breaks in road and rail communications, using temporary bridging equipment;
- Emergency communications (telephone and radio) linking EOCs at various levels of government;
- Help in safeguarding evacuated villages or urban areas; and

- Transportation of emergency supplies including water, food, animal fodder, emergency shelter, etc.

To ensure the availability of support of this kind from military units in emergency, armed forces commanders, at all levels from national to local, must be involved in the processes of disaster-preparedness planning with the civil government counterparts. Not only will such joint planning ensure that each is aware of the requirement and responsibilities of the other but equally important, the constraints and restrictions which may affect the provision of military assistance will be foreseen. In turn, this awareness will—or should—result in various forms of legislation both legal and fiscal to clear the way for effective collaboration between the civil and military authorities. In this process of disaster-preparedness planning, various tasks may be identified that could lead to various degrees of disaster prevention. The construction of retaining walls, flood protection embankments, the deepening or clearance of river channels, and the development of safe evacuation routes, are all tasks which have been undertaken by military engineers in support of national disaster prevention in various countries. Disaster-preparedness planning is likely to include logistic plans for evacuation and emergency support. Logistics is a military activity to which much time is devoted in the training of staff officers and practice in units. Logistic operations planning, as an element of disaster preparedness planning, is a responsibility that can be usefully placed on the military member of any disaster-preparedness planning group. Provision should also be made to ensure that military transport is available to support such logistic operations.

The military can also be usefully involved at the community level in public information and training programs. These activities can include first aid training, simple field engineering relating to providing drinking water, drainage, and field hygiene.

The use of helicopters as a means of reconnaissance, emergency rescue, evacuation, and supply has become a feature of disaster response in many situations. Helicopter operations are extremely expensive and are dependent upon a high degree of ground support. Although it will not be possible to train members of a community to assist in these operations, it is possible to ensure that selected individuals in communities are aware of the requirements for a cleared landing site as well as its dimensions and the necessity to keep this area clear of people and to indicate the wind direction on the ground. These are requirements well understood at all levels in military units and the military responsibility can be to ensure that these are passed on to the civil community.

Food-for-work projects in predominantly agricultural communities can form important elements of disaster-preparedness planning in relation to rehabilitation and recovery. Many food-for-work projects are engineering tasks related to local requirements to create and develop tracks, roads, irrigation and drainage, river crossings, and work of this kind. Although the planning and financing of such work must lie with the community and the civil administration, the employment of military engineers to provide technical skills and equipment, supervision of local labor, and logistic support can be an important element of food-for-work projects.

16. *Regional and International Level*

The successful introduction of foreign—albeit allied and friendly armed forces in disaster relief operations—depends upon careful consideration of the relief tasks, the implication and support needs that such foreign forces may require. Regional associations such as NATO and the Association of Southeast Asian Nations (ASEAN) provide a sound basis upon which planning can take place and examples and experience of such support are available to provide useful guidelines.

Problems concerning immigration, customs clearance, health requirements, logistics support, and costs should be considered and the necessary arrangements made as part of the national disaster–preparedness plan. Even if arrangements and agreements are not finalized at this planing stage, a detailed analysis of the implications and requirements should be made so that a checklist can be prepared of necessary actions when disaster strikes. The national armed forces headquarters is clearly the level of responsibility at which such studies and planning should take place. They should also be responsible for identifying and training liaison officers for attachment to visiting armed forces units.

## *Conclusion*

17. The potential, particularly at the national level, of the armed services to provide well organized, trained, and equipped support to the civil authorities and emergency services is great and should become an element of all disaster-preparedness plans.

*Chapter 9*

# International Disaster Assistance

## Purpose

1. The purpose of this chapter is to outline the main factors which apply to international disaster assistance. The chapter covers the following:

- Modern concept of international disaster assistance;
- Types of assistance;
- Main sources of assistance;
- Arrangements by recipient countries;
- Local problems arising from international assistance; and
- Relationship between donors and recipients.

2. Various references to international disaster assistance have been made in earlier parts of this handbook such as in Chapters 1, 4, 6, and 8.

Where necessary, for ease of handbook utilization, these references are reiterated in this chapter.

## Modern Concept of International Disaster Assistance

3. The concept of international assistance is obviously not new. It dates far back into history and is, perhaps, best exemplified in its earliest forms by military alliances of various kinds. However, its modern era could be said to start with the US Marshall Plan immediately following World War II. This, in fact, constituted international disaster assistance on a massive scale. Europe had been shattered by the disaster of global war. It was clearly going to be threatened by political, economic, and social uncertainties unless it could be given substantial help toward recovery. The Marshall Plan aimed to rehabilitate the economies of the post-World War II European nations to create stable conditions in which democratic institutions could survive. The plan's

concept of economic funding was so successful that it was later extended to underdeveloped countries worldwide.

4. In parallel with these international assistance activities of the US, other programs began to develop. These were mainly concerned with countries that were achieving independence from former colonial powers such as Britain, France, Holland, and Germany. It is interesting, in the context of this chapter, that there was a strong contention by many of these newly independent countries that assistance programs should be implemented on a basis of "aid-without-strings."

5. The significant point about this background of international assistance is that it provides a great deal of experience and precedent on which to develop current and future disaster assistance programs. Indeed, there is no doubt that the modern concept of international disaster assistance has benefited significantly from this previous and more general experience. Thus, today, there is a reasonably clear understanding—by both funding agencies and recipients—what is involved in disaster assistance generally. Moreover, the increasing interdependence of nations tends to give disaster assistance a respectable image and make it an acceptable part of international relations. This is not to say that all disaster assistance programs are without their difficulties; in fact, some of these difficulties are outlined in paragraph 20. However, the fact remains that the overall concept of international disaster assistance is currently recognized by most nations as being valid, practicable, and productive.

6. What should also be recognized (and this is probably the most important factor of all) is that it must be the prerogative of a stricken country to decide whether or not it needs international disaster assistance. To impose or pressure such assistance without full regard for a nation's wishes is to defeat the whole spirit and intention of today's development concept. (See also paragraph 22 concerning donor/recipient relations.)

## Types of International Assistance

7. Brief reference to the use of international resources is made in Chapter 8. The main types of international assistance are amplified in paragraphs 8–11.

### 8. *Pre-Disaster Assistance*

Pre-disaster assistance from international sources can take a variety of forms. The following are typical examples:

*Assistance in prevention/mitigation*

- Assistance in building a system of dams, aimed to prevent flooding.
- Development of monitoring and warning systems; for instance, for volcanic eruption.

*Assistance in preparedness*

- *Planning* – Providing assistance in formulating plans at national and other levels; also with departmental operational guidelines and standard operating procedures.
- *Organization* – Providing assistance in establishing and developing disaster management organizational structures or key points; for instance, the establishment of a national disaster management center, office, or section.
- *Systems and facilities* – Providing or assisting with:
  warning systems;
  communications systems;
  emergency operations centers;
  emergency broadcasting systems; and
  protection of key installations, such as power supplies.
- *Equipment* – Stockpiling of emergency items such as generators, chainsaws, shovels, cooking equipment, shelter materials, and medical equipment.
- *Training* – Providing overseas training; assisting with in-country training.

## 9. *Assistance in Response Operations*

As with pre-disaster circumstances, assistance in response operations can also take various forms. Some common examples are:

- Monitoring and warning of potential disaster impact; for instance, assistance from world meteorological networks or tsunami warning centers.
- Post-impact survey and assessment; for instance, aerial photographic or visual reconnaissance.
- Providing emergency assistance teams; for instance, medical teams, defense force teams, other specialist teams.
- Providing emergency equipment and supplies; for instance, communications, generator, clothing, shelter materials, food, transport and medical supplies.

- Providing specialist personnel; for instance, to install and operate water purification plant.
- Providing temporary major response capabilities; for instance, helicopter capability for various emergency roles (including survey and assessment and food distribution), shipping capability for movement of heavy/bulky supplies, off-road vehicle capability.

### 10. *Assistance in Recovery Programs*

The post-disaster recovery process usually consists of a series of distinct but interrelated programs; for instance, covering infrastructure, medical and health systems, education facilities, and so on. International assistance may therefore be directed toward a specific recovery program or comprise some form of contribution to overall recovery, such as:

- financial grants or credits,
- building materials,
- technical equipment,
- agriculture rehabilitation,
- extended feeding program,
- specialists or specialist teams,
- food-for-work program.

### 11. *Assistance in Future Development*

In many cases, international assistance in post-disaster recovery may develop or merge into long-term development programs; for instance, development of transport systems, agriculture programs.

## Sources of International Assistance

12. Sources of international assistance may vary in the case of individual countries, depending on geographical position, regional groupings, and other factors. It is therefore advisable for each country to make a careful survey of the possible sources that are most suitable and convenient for its own needs.

13. A factor to be taken into account when considering international assistance is the existence, in-country, of various organizations such as the Red Cross, ADRA, and World Vision, because these already have day-to-day international links.

## Arrangements by Recipient Countries

14. One key factor in international assistance is the effectiveness of the planning and preparedness arrangements made by potential recipient countries. These arrangements need to cover a number of aspects, as indicated below.

### 15. *Types of Assistance*

In paragraphs 8–11, likely types of assistance are divided into four broad categories, as follows:

- pre-disaster assistance,
- assistance in response operations,
- assistance in recovery programs, and
- assistance in future development.

Pre-disaster assistance and assistance in future development tend to be of a routine nature and can be processed generally to the same time scale as other international relations. Assistance in response operations and recovery programs usually has a high degree of urgency, which necessitates the application of streamlined arrangements. However, it is suggested that all four categories of assistance should be processed through a common in-country system. This is desirable in the interests of effective coordination, and to adhere to the principle of not changing systems or procedures to deal with crisis situations.

The proposals contained in paragraphs 16–19 have been made with the foregoing considerations in mind.

### 16. *Definition of Responsibility*

The responsibility for dealing with international assistance needs to be clearly defined, especially so that under the pressures of disaster impact there can be no doubt as to who should take relevant action. There are various options for covering this definition of responsibility and two possible options are outlined subsequently. However, it must be emphasized that whatever system is used, it should be clearly set out in the national disaster plan and other relevant documents. It is also important that the chosen system should adequately cover day-to-day liaison as well as urgent response and recovery action.

*Option 1*

*Cabinet Responsibility*
To vest ministerial authority in the Prime Minister but to deal, as required, with matters of special national/international importance.

*Prime Minister*
To be responsible for policy decisions concerning:
- types and scales of assistance required; and
- use of international support services and relief commodities.

*National Disaster Council*
- To be responsible for advising the Prime Minister on matters concerning international assistance.
- To be responsible, during nondisaster times, for ensuring that adequate arrangements are in place concerning international assistance.

*International Assistance Group*
- To be responsible for keeping under review all appropriate aspects of disaster-related assistance and to advise NDC accordingly.

*Option 2*

To include the permanent secretary of the Department of Foreign Affairs as a member of the NDC and process international assistance matters through the Ministry of Foreign Affairs in the normal way.

Clearly, the choice of any option will depend on individual country circumstances, particularly its size. However, it is important to bear in mind that international relations need to be processed from the national level and not lower levels of government.

## 17. *Assistance Channels*

It is also important, in relation to paragraph 16, that the channels for dealing with international assistance should be defined and well understood by both recipients and donors. These arrangements should therefore be clearly stated in the national disaster plan and any bilateral memorandum of understanding that might include disaster assistance.

## 18. *Assessment of Likely Requirements*

In cases where urgent assistance may be needed (as in response and recovery operations), it is helpful if the recipient country has a reasonably accurate idea of the type and scale of assistance requirements. To this end, the following actions by the recipient country should be considered:

- To ensure that an up-to-date inventory exists of emergency supplies and equipment already held in-country.
- To list, based on previous experience and/or the experience of similar countries, the items most likely to be required (tents, generators, etc.) and in what quantities. This, compared with the inventory of existing stocks, should give a very broad guide to what might be required initially for any particular disaster event.
- To ascertain, in general terms, what types of assistance are likely to be available from main funding sources.
- To try to delay a request for funding, or to give a holding reply to funding offers, rather than allow a hurried assistance program to be introduced. This kind of decision or approach is not easy in circumstances where many disruptions and pressures apply. However, the point is worth noting as a desirable guideline.
- To ensure that, as far as possible, items which are known to be unusable under in-country conditions are excluded from requests or offers of assistance.
- To ensure that requests for assistance are made in terms and categories that are clearly understandable to funding organizations and that the correct channels and procedures are used.

## 19. *Reception and Use of International Assistance*

One most damaging aspect of international assistance is inefficiency by the recipient country in arranging for reception and use of resources and commodities. For instance:

- If valuable supplies cannot be properly stored and therefore are left on wharves to deteriorate under conditions of post-cyclonic rain, or similar circumstances.
- If helicopter assistance units cannot be fully used because of insufficient survey and assessment information for their tasking.
- If the scale of assistance requested and accepted is beyond the capacity of the recipient country:
  - to process through ports, airfields, or overland routes; and/or
  - to distribute effectively, perhaps due to lack of transport or inadequate transport routes.

It is therefore most important that these issues should have been carefully considered during the planning process and appropriate arrangements laid down in relevant plans. In this regard, it is recognized that disaster itself (e.g., secondary earthquake shocks) may disrupt planned arrangements but, clearly, this kind of contingency has to be accepted by both recipient and funding agency or source alike.

## Local Problems Arising from International Assistance

20. In some circumstances, problems can arise locally from international assistance, albeit indirectly. Some examples drawn from case studies are given below:

- *Overdependence on assistance*
  In some countries, it has been found that overdependence on assistance can develop. For instance, traditional custom in one country has been for communities and individuals to store food items in readiness for the cyclone season. However, over a period of years, the government—using international as well as national resources—has quickly responded to post-disaster needs by liberally distributing food. Consequently, the former tradition of self-reliance has been eroded and overdependence on assistance has taken its place.

- *Effects on local economy*
  Rapid injection of assistance, especially food items, can upset a local economy. This particularly applies when local markets and rural production are interdependent. However, governments are often in a dilemma immediately post-impact for a number of reasons, including the following:

  - Often, local food stocks held in stores and warehouses may be damaged by the disaster; or bulk refrigerated goods may face deterioration through lack of power supply. To minimize food wastage, therefore, a government may sometimes be forced to buy up local stocks and distribute them as food assistance. This can upset local markets, even before international assistance appears on the scene.

  - International assistance—prompted by the best of humanitarian reasons—may arrive very early after impact, while local food stocks are still adequate. This can embarrass the local situation but the recipient government, in the interest of good relations, is not in a position to reject such assistance.

The effects on a local economy can usually be rectified in the longer term. However, in immediate post-impact conditions, they can add considerably to the problems and pressures. Moreover, hardship is often imposed on rural food producers as they struggle to rehabilitate their business and reestablish former market outlets.

- *Oversupply of assistance*

  Oversupply of assistance is another well-known problem area which can particularly apply to severe and widely publicized disasters. It may result in assistance of all sorts, shapes, and forms being showered on a stricken nation, with little or no regard for its suitability or the amount of assistance already received.

  In one recorded case, a large supply of yellow bikinis was sent to refugees trying to subsist in semi-arctic conditions. In another case, supplies of high-heeled shoes were sent to victims who were never likely to wear them. In a third case, a well-meaning overseas community collected a huge amount of fruits and had it flown by chartered aircraft to a neighboring country. On arrival, the fruits had to be destroyed because of the danger of introducing fruitfly and thus risking the future of indigenous crops.

21. It must be emphasized that the examples quoted in the foregoing paragraph are in no way intended as a criticism of the international assistance principle. They are included to serve as a reminder of the very real problems which can arise, particularly in the unsettled circumstances which apply following disaster impact. Indeed, the encouraging signs are that the general situation is steadily improving, A great deal of experience has now been accrued by a wide range of funding countries and organizations so that, hopefully, the local problems of recipient nations will be kept to reasonable levels.

## The Relationship between Funding Sources and Recipients

22. There is no doubt that the success of international assistance efforts depends very significantly on the understanding between assistance agencies and recipient nations. This particularly applies to mutual preparedness arrangements. In the past, there have certainly been instances where better donor–recipient preparedness would have eliminated many problems.

In fact, most of the major problems in international assistance can be avoided if the following factors are recognized:

- Assistance agencies need to exercise a sensitive approach. When assistance is needed, the recipient nation is usually in some form of post-impact shock. The shock period can extend well into the recovery phase and is particularly illustrated by difficulty in identifying assistance needs.

- Assistance agencies also need to be patient. From the recipient viewpoint, there is nothing more aggravating and counterproductive than impatient overkill in assistance offers.

- The practical preparedness measures of assistance agencies are also important. An understanding of likely in-country circumstances, of logistical considerations, and of local conditions that affect receipt and use of assistance commodities, is a key factor in successful assistance.

- Optimum coordination between government and nongovernment agencies is strongly desirable, from both the in-country and overseas viewpoints.

Overall, the desirable concept is one of a tactful mutual relationship throughout the whole process of preparedness, response, and recovery. In this way, when assistance needs to be applied, it is merely one phase of an ongoing dialogue, rather than a sudden shock injection into an already traumatized recipient country.

*Chapter 10*

# Leadership in Disaster

---

## Purpose

1. The purpose of this chapter is to outline the main considerations which apply to leadership in disaster. The chapter covers:

- leadership under conditions of crisis;
- some ramifications of political leadership;
- leadership in the direction of response operations;
- leadership in resource organizations;
- community leadership; and
- some leadership attributes and desirabilities.

## Leadership under Crisis Conditions

2. In all nations and all societies, leadership has a prominent and powerful role. It influences all aspects of life through all levels. If we do not like our leaders, or if we grow tired of them, we make changes. Mostly we do this by peaceful means. Sometimes, however, the change is violent.

3. Just as leaders may change, so do leadership needs and requirements. This usually happens because societies are constantly adjusting to change and development. Under stable conditions (that is, when a nation is not affected by war, civil unrest, disaster, or other forms of disruption), although the tasks of leaders may be onerous, they mostly tend to be orderly and reasonably straightforward. There is usually time to make calculated decisions based on methodical assessment of the various factors involved. Indeed, quite often, decisions are made only after a process of consideration and reconsideration, of drafting and redrafting. Also, there is normally a well-founded organizational system within which this process takes place, plus a framework of legislation to provide formalized support and confirmation. Even so, in spite of stable conditions, some leaders have to operate under unpleasant and stressful circumstances. For instance, leaders of firefighting teams carrying out what for them are normal duties, often have to accept high risks themselves and make decisions that affect the safety and lives of others.

4. By contrast, under unstable and disruptive conditions such as those which apply in disaster, the tasks of leaders usually become more difficult. This applies at most, if not all levels, from government ministers through to village leaders. Some of the factors which may affect leadership during disaster conditions are:

- loss of some designated leaders, who may be put out of action by the disaster itself;
- failure of designated leaders to cope in traumatic conditions with which they are not experienced;
- reduced effectiveness of subordinates because of various crisis effects;
- lack of information on which to make decisions; or inadequate facilities for presenting information to leaders/decision makers;
- lack of or disruption to communications, which inhibits the ability of leaders to consult with other key persons, or to convey instructions rapidly and clearly;
- severity of post-impact conditions which may make it difficult for leaders to determine courses of action and priorities; and
- loss or delayed availability of resources in terms of personnel, equipment, transport, relief commodities, and so on.

There can, of course, be many other factors which affect leadership under disaster circumstances, depending on specific situations. However, those mentioned above serve to illustrate this particular point.

5. What is equally or perhaps even more significant is that the importance of decision making at various levels may vary from that which normally applies. In fact, in many instances, the relative importance of decision-making levels may be reversed. This can apply particularly in the early post-impact stages of a disaster when decisions taken at the "disaster front" (i.e., within the stricken area) may be more crucial than those taken at national provincial levels. Enlarging upon this point, some or all of the following considerations may apply:

- Prudent decision on precautionary measures during a pre-impact warning period, taken by leaders at, say, municipal or village level, may ensure that communities are in safe havens or protected areas when disaster impact occurs. This obviously mitigates disaster effect.
- Decisions by those same leaders concerning post-impact priorities for rescue, temporary evacuation, or other action may be crucial to the lives and livelihood of large numbers of people.
- Decisions to organize and implement self-help measures—pending assistance from outside—may significantly defer or alleviate potential

hardship for those who have lost their home and/or means of sustenance.

- The tangible strength of leadership shown by local leaders through personal example, courage, and demeanor may be far more significant to large numbers of disaster-affected people than, for example, stirring television or radio messages sent by senior government ministers who are in areas of relative safety.
- If there is undue delay (real or perceived) in mobilizing in-country resources to relieve a stricken area, then the credibility of leadership at higher levels of government is likely to come under question.
- Similarly, if the national government is seen as not making best possible use of overseas assistance, its future right to leadership may be challenged politically.

6. It is clear therefore that conditions of disaster impose many unusual pressures and demands on leaders at all levels. In disaster management terms, this clearly indicates the need to analyze and study leadership requirements and, where possible, to attune and train key persons accordingly.

## Some Ramifications of Political Leadership

7. Certain aspects that affect political leadership have already been aforementioned. However, the policies and attitudes adopted by political leaders may have far-reaching ramifications for disaster management and, in particular, for response operations and recovery programs. These effects can be illustrated by considering circumstances prior to, during, and after disaster impact.

### 8. *Pre-Disaster Considerations*

Regrettably, it has to be said that political leadership can expose whole nations to the threat of disaster. There are many reasons for this, some of which already been referred to in Chapter 3 on national disaster management policy. However, within the specific context of this current chapter, some examples of the ramifications of political leadership are given below:

- The apathy of political leaders to the disaster aspect of national affairs can be significant. This can occur for a variety of reasons and sometimes amounts to a form of calculated risk-taking. However, when it does occur, the whole capability of a nation to cope with disaster is seriously or sometimes dangerously downgraded.
- The threat of disaster and the true significance of disaster effects may be underrated by political leaders. This may particularly

apply if the disaster threat is low key and spasmodic, and if no consideration has been given to risk analysis and vulnerability assessment.

- There may be overreliance by political leaders on the willingness and capacity of friendly countries and international organizations to provide rapid assistance if disaster strikes.
- Perhaps in consequence of some or all the foregoing factors, no in-country organization has been set up to monitor and advise on disaster-related affairs. Thus, overall national awareness is low and preparedness virtually inexistent.

In sum, the effects of inadequate political leadership on these pre-disaster requirements is to produce a nation that is inadequately prepared to cope with disaster. People therefore suffer unduly and national assets and development are seriously jeopardized.

## 9. *Considerations during Disaster*

Immediately prior to and following disaster impact—during what is usually called the emergency phase—political leaders can decide in the spur of the moment, which may produce far-reaching results, some of them counterproductive. The following examples, drawn from factual circumstance, serve to underline this contention:

- A senior minister, acting in the absence of his prime minister, completely ignored the existence of national and regional disaster plans, even though these plans had proved effective in an earlier disaster. He set up a dubious ad hoc organization at national level and attempted to run all counter-disaster activities from this level, disregarding established government organizations at lower levels. It took approximately seven precious days to disentangle this ad hoc organization and replace it by the more orthodox one intended under existing plans. The result was that people suffered unnecessarily due to delay in relief operations.
- In contrast, on another occasion, the reversal of an original ministerial decision markedly improved response action. This case involved the establishment of a standard-type national emergency operations center (NEOC) to replace a somewhat loose committee system. The great advantage of the NEOC was that it provided a much-needed "information picture" on which to base decisions and priorities.
- During a complicated relief operations in the wake of a major disaster, a senior politician openly abused a very experienced disaster management official, accusing him and his colleagues of

gross incompetence. The politician had no disaster management experience and knew nothing of the complexity and extent of the relief operation. He was merely concerned to illustrate to his political supporters that he was shaking up the system, supposedly on their behalf. His ill-informed efforts achieved nothing apart from wasting the time of officials who had to make important decisions and to fulfill critical relief tasks.

- In another instance, a prime minister ruled that government ministers were not be involved in the post-disaster distribution of relief supplies. He particularly banned them from participating in helicopter supply missions. His action was aimed at ensuring that no political pressures were brought to bear that might lead to unequal and unfair distribution of assistance.
- One extreme case is, hopefully, a rarity. Priority in relief assistance was given to members of the ruling political party.

It would, of course, be inaccurate to infer that political interests in emergency operations are ill-considered and wrongly motivated. Disaster raises many aspects of political involvement and, indeed, affected communities can justifiably expect that their elected representatives will look after the peoples' interests in time of need. The important thing from a disaster management viewpoint is that there should be recognition of the fact that political intervention is highly likely to occur in varying forms; and that disaster managers—especially those with senior responsibilities—should be aware of this and be prepared to handle the resultant problems.

## 10. *Post-Disaster Considerations*

In post-disaster recovery and associated development programs, political leadership clearly plays a primary role. In this context, decision making tends to have very profound long-term effects. Ideally such decision making normally needs to involve:

- an optimum data base, including relevant factors from the current disaster and from previous ones, plus projections for feasible future development;
- clear definition of post-disaster recovery and development strategy;
- accurate determination of individual programs within such strategy;
- clear definition of responsibility for implementing the strategy and its component programs;
- effective management and coordination of programs/projects; and
- relevant adjustments, where applicable, to current and future national development plans.

Unfortunately, it is not always possible to adhere to the foregoing desirabilities. Factors which affect this may include the following:

- recording of relevant data during the emergency phase has not been adequate;*
- post-disaster review has not been effectively carried out,* [* Both these items may be crucial for providing an adequate data base.];
- special committees or commissions set up to review the disaster, or selected aspects of it, may produce findings that are politically unpalatable. Such findings are therefore conveniently disregarded or given low priority. A case is known where a government, having set up a special disaster review commission, used the relevant findings to give the impression that it (the government) had already taken preemptive action on the many shortcomings revealed.
- Funding sources of international assistance may pressure an affected government into premature decisions. This may arise from budgeting, programming, or similar requirements on the funding sources' side.
- The timing of recovery programming may be such that the government—for reasons of retaining political power—needs to be seen as acting vigorously and quickly. This may lead to programs being prematurely approved, thus forfeiting some of their effectiveness.

11. The considerations outlined above clearly indicate the importance of political leadership and decision making in relation to the total spectrum of disaster management. One extremely important corollary is that national leaders and politicians at all levels need to be kept aware of disaster problems and of the necessary countermeasures. As with other aspects of government, it is the responsibility of senior officials to ensure that this awareness is maintained. In turn, this requirement underlines yet again the need for some kind of disaster management section within a government that can monitor policies and programs.

## Leadership in the Direction of Response Operations

12. As indicated in Chapter 2, the effects of disaster impact can be wide-ranging and traumatic. This means that response operations usually have to be carried out under conditions that place heavy personal demands on both victims and disaster management personnel. Thus, the direction of such operations obviously calls for a clear and decisive leadership. However, it is often difficult to achieve effective response leadership for reasons such as those outlined on the next page.

## 13. *The Problem of Operational Coherency*

It is often difficult for senior disaster management officials (e.g., those responsible for the direction of response operations) to maintain and implement a coherent operational plan of action. For instance, in post-cyclonic conditions of very low cloud and heavy rain, it may be impossible to carry out vital air reconnaissance. Similarly, a very fast-moving wildfire (bushfire) front may make it extremely difficult to mobilize and apply resource organizations to full effect. Response leaders need to be aware of possible problems in operational coherency and to recognize the limitations these can impose, albeit temporarily, on the effectiveness of operations.

## 14. *Access and Movement*

Closely linked to the problem of operational coherency is that of access to and movement within a disaster area. This can apply to various circumstances, from a highly toxic industrial disaster to a post-cyclonic mudslide/landslide. Airborne resources can solve some of the problems inherent in these situations.

## 15. *Quality of Resource Organizations*

(See also paragraph 19)

The effective overall leadership of response operations can also be influenced by the quality of the resource organizations involved. If these organizations are highly skilled, self-contained in logistics and communications, and experienced in the type of operations being undertaken, then the task of overall leaders/coordinators tends to be simplified. This can be a significant factor under most circumstances but especially when the response task is extensive and complicated.

## 16. *Problems of Coordination*

A frequent problem in the direction of response operations is coordinating the various activities undertaken by resource organizations. (This problem is also referred to in Chapters 4 and 17). Wide international experience tends to indicate that there is no easy solution to this problem. However, from a leader's viewpoint, it is important to understand that success or failure in achieving coordinated effort by resource organizations will depend significantly on the following:

- capability, experience, and personal qualities of the coordinating leader(s);

- quality of leaders of resource organizations;
- professional competence of individual organizations;
- previous experience of these organizations in dealing with disaster problems, especially in conjunction with other organizations;
- relationships between the coordinating leader(s) and leaders of individual resource organizations, and between individual leaders themselves;
- practice in coordination during exercises; and
- clear allocation of roles to individual organizations and adherence to such roles, in terms of both tasking and implementation.

### 17. *Delegation of Leadership Responsibilities*

Another common shortcoming in response leadership is failure or inability to delegate responsibility. Because events immediately post-impact are complicated, urgent and pressurized, many leaders seem to think that they "must see the crisis through." Almost inevitably, this leads to overwork, strain, and fatigue, in consequence of which the capability, competence and, especially, judgment of leaders become seriously downgraded. In other words, there is absolutely no point in leaders carrying on until they are "dead on their feet." Therefore, it is prudent, wherever possible, to direct response operations as a team (e.g., an operations control group of three members) that allows for adequate sharing of leadership responsibilities and pressures.

18. In sum, effective leadership of response operations depends greatly on recognizing the types of problems likely to arise and endeavoring to offset them by prudent management and preparedness. In this regard, a valuable guideline for overall leaders/coordinators is to task resource organizations accurately and then leave the organizations to complete their tasks using their own systems and procedures.

## Leadership in Resource Organizations

19. As indicated in paragraph 15, high-quality resource organizations represent a considerable asset in overall response capability. In this respect, leadership of individual organizations is obviously an important consideration. However, perhaps the only common factor which can be said to apply to the leadership of resource organizations is that its requirements can vary greatly. There are several reasons for this, including the following:

- The nature of the organizations; some are professional and full-time, while others are voluntary and part time.

- The different tasks undertaken by resource organizations tend to require different leadership emphasis and characteristics. For example, a debris-clearance or rescue team usually requires a vigorous, physically tough type of approach. By contrast, the essence of approach and action by a welfare team is more oriented toward care and comforting; yet these teams may have to face considerable strain transmitted from distressed and traumatized victims.
- Changes in leadership and/or membership of resource organizations.
- Changes in capability of organizations because of various factors such as loss of experienced team members, or temporary allocation of organizations to unfamiliar tasks.
- If the disaster threat is spasmodic and low level, it may be difficult to maintain enthusiasm and skill levels, especially in voluntary organizations.

There are therefore many demands and challenges facing leaders of resource organizations if the ideal counter-disaster requirements is to be met. This requirement is that the overall response leader/coordinator can task individual resource organizations in the knowledge that, under their respective leaders, tasks will be completed competently and expeditiously.

## Community Leadership

20. Leadership at community or local level is another very significant factor in disaster management. It is suggested that it is particularly important for disaster management officials, at all levels, to have a good understanding of leadership at this level because of its "disaster front" ramifications.

21. Some important factors that affect community leadership are as follows:

- As mentioned on several occasions previously, the nature, severity, and frequency of the disaster threat will strongly condition community attitudes and leadership. An aware, alert, and informed community constitutes an enormously strong asset for coping with disaster, particularly since such a community will respond to elected leadership.
- In times of disaster, communities tend to turn to established and traditional sources of leadership. The sources may comprise local councils and similar organizations or individuals who are recognized as holding hierarchical leadership positions. This aspect also underlines the fact that it is not prudent to try to change established community systems to deal with disaster.
- The attitudes of higher levels of government may also affect community leadership. If a government is unduly paternalistic and

its control reaches right down into the roots of the community, that community may lose initiative and self-reliance. Therefore, when it is perforce left to its own devices under disaster conditions, it may not respond effectively to local leadership. In this regard, it must be stated that in most disaster-prone countries nowadays the development of self-reliance at community level, and the importance of good local leadership, are well recognized as desirable disaster management objectives.

22. It is also at community level that the influence of so-called "emergent leaders" is usually most felt. Such leaders, as the title implies, tend to be persons who stand out from within the community, or perhaps from within the lower echelons of organizations, and take the lead under crisis conditions. It is difficult to generalize on the circumstances in which emergent leaders will assert their influence but obviously this is likely to happen when, for example:

- established leaders are put out of action, or
- groups of people are cut off from the rest of the community, or
- the recognized leadership system fails to respond adequately to a particular set of disaster conditions.

There are, of course, both advantages and disadvantages where emergent leadership is concerned. Sometimes it can be highly beneficial in helping resolve problems at community level. At other times, it may exacerbate already difficult situations through unreasonable criticism and "splinter" action.

For disaster managers generally, therefore, the emergent leader phenomenon is one which needs to be recognized and dealt with as best possible in relation to relevant circumstances.

23. In many cases, it has been found that legislation can help considerably in community leadership. By clearly laying down the responsibilities of local government and of the various counter-disaster organizations, such legislation reduces the possibility of confusion concerning community leadership and, over time, helps develop an accepted leadership structure.

## Some Leadership Attributes and Desirabilities

24. As will be recognized from the foregoing paragraphs, there are many different aspects, conditions, and factors that influence leadership. Thus, when it comes to analyzing leadership—especially in terms of attributes and

desirabilities, many variations can apply. It follows that what is contained in this part of the chapter can therefore be no more than a basis of consideration. Ultimately, it must be up to individual disaster managers to adapt leadership principles, guidelines, and experience to their own circumstances.

25. Perhaps a start can be made by making two basic suggestions:

- First, it is not possible to "learn" leadership by memorizing and applying certain facts. We cannot, for example, "learn" courage, which most people would regard as a worthy and useful leadership quality.
- Second, only very few people can be classified as ideal, all-around leaders. At the same time, however, most of us do have some useful leadership qualities.

If one accepts that the above suggestions are reasonable, it tends to indicate that there is limited merit in trying to judge leaders or potential leaders in terms of personal qualities alone. The latter process is sometimes done against a list of desirabilities such as courage, dedication, enthusiasm, sense of humour, endurance, cheerfulness, and so on. This sort of process is bound to end in disillusionment, since few if any individuals are likely to meet such a specification. Therefore, while personal qualities undoubtedly play an important part in leadership capability, it seems prudent to look for other aspects as well. The following might therefore form a possible combination of desirable objectives for leadership in disaster management.

- *Personal qualities*
  To try to identify one's own personal leadership qualities and, where possible, develop and expand them to best capability.
- *Professional competence*
  To aim to develop high standards of knowledge, skill, and ability appropriate to tasks and circumstances. Essentially—and especially applicable to the disaster management role—this competence means knowing what to do and how to do it. Such competence is highly important because it inspires respect and trust in the leader.
- *Self-confidence*
  In many ways, though not necessarily entirely, self-confidence stems from professional competence; because someone who is competent in a task must be justifiably confident to carry out that task. Again, this reflects on the leader's status and stature.
- *Sound judgement*
  This tends naturally to develop from the combination of professional competence and self-confidence. It is an enormously

strong leadership attribute because it brings out and consolidates trust from the team or organization. A leader with sound judgement is not likely to jeopardize colleagues.

- *Accurate decision making*
  This, again, tends to be an outcome of foregoing qualities. It is obviously critical in committing a team or organization to a task. Moreover, as with sound judgement, it reflects on the relationship between the leader and the team particularly in terms of trust.
- *Ability to communicate*
  The ability of a leader to communicate clearly and accurately with colleagues is another key leadership requirement. This applies not only when a team or organization is being committed to demanding and perhaps hazardous tasks; it also helps the leader transmit a sense of purpose and integrity.
- *Personal example*
  It is important that the leader must set a personal example when this is required. There is little point in a leader having attributes such as those listed above unless he/she can play a part in tasks, when this becomes necessary, and to be seen as being able to do so.
- *Appropriate style*
  Different tasks tend to require different leadership emphasis and characteristics. The adoption by a leader of an appropriate leadership style is obviously necessary especially because it reconciles that person with both the task and the team/organization. (Some brief comments on leadership styles are contained in paragraph 26.)

## 26. *Leadership Styles*

In examining leadership styles, gradations are sometimes used to illustrate variations that can apply between strong leader influence and strong team influence (sometimes called "boss-centered" and "subordinate-centered" influence). One example of this is given below.

*Tells* – The leader decides, tells the team, then expects the orders to be carried out without question.

*Sells* – The leader still decides, but recognizes some possibility of resistance and, therefore, takes care to "sell" the ideas to the team.

*Consults* – The leader asks for views from the team before deciding but the decisions are still those of the leader.

*Joins* – The leader puts the problem to the team and after discussion takes the majority decision.

This type of gradation has sometimes been criticized on the grounds that a leader is there to lead and that, therefore, team influence should be only minimal. However, a case is known of a highly successful volunteer mountain rescue team in which all members were trained and qualified in the leadership role. The reason for this was that calls on the team were usually very urgent and—being volunteers—not all members were always immediately available. Therefore, any member had the capability of mobilizing the team and, if the leader or deputy were unavailable, could initiate a rescue mission. The team's record of success clearly underlined the wisdom of this "multileader" system. It would seem, therefore, that disaster management authorities would be well advised to take an analytical approach to this and similar aspects of leadership.

## 27. *Leadership and Training*

Another useful factor in the development of disaster management leadership is the strong and positive link between leadership and training. If a team or organization is trained to very high standards, it will normally develop a high degree of professionalism. This, in turn, promotes an attitude of pride and strong esprit de corps. Leadership of such a team or organization does not usually present many problems.

*Chapter 11*

# Organization

## Purpose

1. The purpose of this chapter is to consider major aspects of disaster management organization.

2. The subject of organization has already been referred to in Chapter 4 on major requirements for coping with disaster.

3. Aspects covered in this chapter are:

- factors affecting disaster management organizational systems;
- organizational components, including emergency operations centers; and
- organizational structures.

## Factors Affecting Organizational Systems

4. Certain factors that influence the establishment of organizational systems are set out in paragraphs 5–15 below.

5. ***A Key Objective***

A key objective for any counter-disaster organization is that it should be capable of intervening rapidly and effectively in disaster situations.

6. ***The Disaster Threat***

This is a fundamental consideration that basically determines the shape and size of the organizational system. Factors that need to be considered are the following:

- nature of the threat (in terms of natural disaster, man-made disaster, civil unrest, etc.);
- degree of the threat (whether it is severe, moderate, or slight); and
- pattern of the threat (in terms of frequency and extent).

## 7. *Government Policy*

Government policy in relation to the threat will invariably have a strong influence on the broad national organizational framework. Such policy should be stated in one or all of the following documents:

- national disaster management policy statement;
- disaster legislation;
- national disaster plan; and
- some form of policy instruction concerning the establishment of a counter-disaster organization.

It would be unusual for disaster management officials to embark on the establishment of a counter-disaster organization without clear policy direction from the government.

It may be worth noting that in many cases, in practice, organizational structures have evolved from some form of planning base (e.g., a national plan, or even a number of provincial plans) in the absence of a national policy statement or disaster legislation. However, the establishment of an organizational system is undoubtedly more straightforward if there exists a policy statement and/or legislation.

It is also worth noting that modifications to the national organization pattern may be necessary at lower levels. There, modifications may arise because, for instance, some province/regions may be more disaster-prone than others, or for reasons of resource limitations, two or more government areas (e.g., municipalities) may have to combine to form a viable organizational entity.

The examples of existing disaster legislation, contained in Chapter 7, give some interesting insights into the interrelationship between government policy and organization.

## 8. *Operational Requirements*

The type of operations required, in response to the threat, will also have a direct bearing on the organizational system. For instance, in very flood-prone areas there is often a requirement to establish special water-borne resources (e.g., flood-boats for rescue and relief purposes). This, in turn, may require special organizational arrangements (including specialist staffing and training). However, though such arrangements may require organizational modifications, they do not usually affect basic principles and structures.

## 9. *Availability of Resources*

Whatever organizational framework is considered, it must fit the resources that are likely to be available in terms of resource organizations (government and nongovernment), facilities, equipment, supplies, and personnel. An organizational framework that is too cumbersome and/or top-heavy will obviously not achieve optimum effectiveness. The same applies if the framework is too small to handle the resources involved.

## 10. *Definition of Responsibilities and Functions*

A necessary prerequisite to the framing of a disaster management organization is that there should be clear definition of the responsibilities that the organization is required to cover and the functions it is to perform.

## 11. *Use of Existing Structures and Organizations*

As emphasized in Chapter 4, another important consideration is that the disaster management system should be an adjunct to the existing government structure. It should also be capable of maximizing the use of existing organizations (both government and nongovernment) for disaster-related purposes.

## 12. *Requirements for Direction and Management*

The organizational system obviously needs to include arrangements for its own direction and management. This aspect is also relevant to paragraphs 30–35 below, which are concerned with organizational structures. However, fundamental considerations are that there should normally be adequate provision for:

- policy direction from government;
- ministerial responsibility;
- a central body, responsible for implementing government policy and ensuring effective management during both nondisaster times and during a disaster itself. This usually consists of some form of national disaster council or national disaster committee;
- a small group responsible to the central body which, during response operations, undertakes the tasking of resource organizations, thus leaving the central body free to make policy decisions and determine priorities; and
- relevant management arrangements within the government structure at lower levels that harmonize with arrangements made at the national level.

- Appropriate arrangements throughout the total system for handling day-to-day disaster management affairs.
- Optimum coordination of counter-disaster effort.
- Community involvement and participation.

## 13. *Incorporation of Organizational Components*

All counter-disaster organizational systems need to contain and operate a number of different components, covering aspects such as warning, information management, and so on. The system therefore needs to be capable of incorporating these components covered in more detail in paragraphs 16–29.

## 14. *Some Compromise Considerations*

In many countries, financial and other limitations may make it difficult to have a separate specialist organization or department for counter-disaster purposes. It is more likely that counter-disaster requirements will have to be met by some system of departmental compromise, such as attaching a counter-disaster section to an appropriate government department and using government (and nongovernment) resources in an optimum counter-disaster role.

Certain compromise considerations also tend to apply to the use of personnel. These indicate that whatever the circumstances, counter-disaster organization should be kept as straightforward as possible. There are two reasons for this:

- First, some parts of the organization (such as those primarily concerned with response operations) will be activated only when required. Thus, the simpler they are, the more easily they can be activated and the more readily the various functions can be remembered and performed by the persons concerned.
- Second, many of the personnel involved will have to be brought into the organization from other departments and agencies. Therefore, there is merit in ensuring that functions are kept as straightforward as possible.

## 15. *Organizational Stress Factors*

Any counter-disaster organization must obviously be capable of coping with the stresses and pressures of disaster events. It is therefore important that its components (e.g., communications, warning system, and facilities) should function to the highest possible levels of efficiency and that they should be safeguarded, as far as possible, from disaster effects. It is also important that government and NGOs, most of which necessarily have to extend from their normal role into a counter-disaster role, should be organized so that they can cope with the additional disruptions, stresses, and pressures involved.

## Organizational Components

16. The following components are likely to be necessary for the majority of counter-disaster organizations, though other components may need to be included in specific circumstances. The components are listed in brief format only to provide a reasonably concise checklist for disaster management officials who may be tasked with establishing or reviewing a disaster-related organization.

### 17. *Facilities*

Various facilities are needed to house emergency operations centers, communications sections, conference/briefing rooms, offices, equipment stores and so on. In many countries, a dual-purpose approach is taken toward this aspect. For instance, suitable office space can be speedily converted for use as an emergency operations center by having maps and skeleton display boards readily available (for presenting situation information, resources available, tasks with priorities, tasks being undertaken, etc.). Similarly, appropriate temporary conversion for disaster purposes can be made of police headquarters or military headquarters facilities. Obviously, however, custom-built facilities are preferable, provided their establishment and use is cost-effective.

### 18. *Equipment and Supplies*

Most disaster management organizations need to hold various categories of emergency equipment and supplies. These will vary according to individual requirements but usually the following need to be stored and maintained:

- vehicles;
- emergency equipment such as generators, emergency lighting sets, picks, shovels, tarpaulins, blankets, tents, and communications equipment; and
- training equipment.

## 19. *Trained Staff*

Adequately trained staff must be available to perform all the functions that have been allotted to the disaster management organization. As indicated in paragraph 14, many of the staff involved in counter-disaster activities will necessarily have a part-time role only. Therefore, if the organization is to achieve the key objective of being able to intervene rapidly and effectively in disaster situations, it needs to be capable of ensuring that adequate training programs are maintained.

## 20. *Communications*

Without adequate and effective communications, no counter-disaster organization is likely to function satisfactorily. As a notable military commander is reputed to have said, "Without communications, all I command is a mahogany desk. A mahogany desk is not an effective weapon of war." In disaster management terms, four particularly important considerations apply to communications systems, in that:

- They must provide adequate facilities for the normal day-to-day functioning of the organization.
- They must be capable of extending from the day-to-day role into the wider and more demanding functions of response operations.
- They must, when necessary, provide a mobile capability.
- They must have adequate reserve or back-up capacity to meet emergency demands.

Sometimes, because of financial and other constraints, it is not possible to provide a special communications facility to fulfill the above needs. Therefore, this may mean using the most proficient available network (e.g., a police communications system) and supporting this with other networks for backup or standby emergency purposes. Existing regular broadcasting systems are also invaluable for disseminating disaster-related information.

It is usual and effective to coordinate the capabilities of available communications networks under relevant disaster plans, so that these can be readily called upon by disaster management officials as required. (See also Chapter 12 on plans).

## 21. *Warning Systems*

The effectiveness of warning systems is one of the most critical components in any nation's capability to deal with disaster. This requirement therefore needs to be adequately covered within all organizational systems at all levels. The following are key requirements in warning systems:

- *Capability to receive international warning*
  For example, warning of developing cyclones under international meteorological arrangements or tsunami warnings from designated international centers.
- *Capability to initiate in-country warning*
  For example, in cases such as volcanic eruption, flood, landslide, fire, and hazardous chemical incident.
- *Capability to transmit warning from national and other levels*
  For example, by radio broadcast, television, or special warning systems.
- *Capability to disseminate warning at local community level*
  For example, by loud hailer, siren, warning pennant, bells, or messenger if circumstances so demand.
- *Capability to receive warning and act upon it*
  This requires possession of or access to a radio/television receiver, being in hearing/seeing distance of signals, knowing what the various warnings mean, and knowing what to do.

If there are limitations or disruptions to any of these capabilities, warning is unlikely to be totally effective. Most limitations or disruptions are caused by disaster events themselves; for example, the inability of a radio broadcast system to function due to cyclone, earthquake, or other damage.

In situations where there is no advance warning (e.g., earthquake or explosion), the counter-disaster organization is at a disadvantage in terms of readiness to respond. Alerting and activation arrangements that are contained in relevant plans and operating procedures need to cater for this eventuality as far as possible.

## 22. *Information Management*

The importance of information in coping with disaster situations cannot be overemphasized. Even the best possible organizational system, manned by the most professional and expert staff, is virtually useless without relevant information and the means of processing it.

The quality of information is also of key importance. Information must be accurate and it must be up-to-date. Otherwise, response will not be timely and it will not be directed to critical areas in the right order of priority.

Sources of information also need to be considered carefully to provide two main information categories, namely:

- *Crisis information*– the various forms of information that apply directly to a particular disaster situation. This includes weather reports, damage reports, situation updates, and so on. It is dynamic information, directly related to the ongoing event.
- *Background information*– records of previous disasters, map information, census information, and so on. This is static information but, in its way, it may be just as important as crisis information.

Organizationally, therefore, the need is to ensure that adequate arrangements exist for the acquisition of crisis information and that background information can be readily accessed for use as required.

Information processing tends invariably to be geared to the simple disaster information cycle of:

- information,
- assessment (or evaluation),
- decision making, and
- dissemination of information and decisions.

It is also important that the information processing system ensures that information is distributed not only vertically through the organization but horizontally as well. This simple point may sometimes be overlooked.

For effective operation of the disaster information cycle, certain organizational facilities and capabilities are necessary. These include:

- emergency operations center (static and mobile);
- good communications;
- capability for survey and assessment;
- facilities to acquire, transmit, receive, display, collate, assess, store, and generally handle information;
- provision for presenting information in such a way that it can facilitate decision making; and
- trained staff to operate the information management system.

Overall, in considering relevant organizational aspects, it is important to bear in mind that the main purpose of information acquisition and processing is that it must enable disaster management officials to identify, prioritize, and allocate response tasks.

## 23. *Liaison Facilities*

Most organizational systems need to include facilities for liaison with various interrelated agencies. The precise nature of these facilities and the channels used will usually need to be decided based on individual situations. However, possible examples are:

- *Liaison necessary to incorporate NGOs into a national counter-disaster system*
  Many governments prefer to effect liaison with NGOs on a group basis where NGOs form themselves into some form of support organization and, in principle, liaise with government, as a group, through defined channels. For day-to-day matters, this could be through an NDMO; and during disaster operations, through a central government tasking group. Channels for this type of liaison obviously need to be clearly defined in relevant plans.
- *Liaison necessary for requesting and providing international assistance*
  Channels for this liaison, again, need to be clearly defined if international assistance is to be timely and effective. Liaison needs to be ongoing in nondisaster times as well as during disaster events. Usually, this type of liaison involves the Department of Foreign Affairs but it is important that the NDC or its equivalent can establish and maintain clear and rapid contact with sources of assistance, as and when necessary. As with the NGO liaison quoted above, arrangements need to be outlined in relevant plans.

## 24. *Media and Public Relations*

Organizational facilities are also required to ensure a sound relationship with the media and for public relations generally. There are two important aspects of relationship with the media.

- First, the media needs information for its own news purposes and, in times of disaster, it is usually in everyone's interest that such information should be accurate. Otherwise, all sorts of repercussions can arise, some of which may exacerbate an already disruptive situation.
- Second, the media can be a valuable disaster management resource, given its capability to disseminate official information and guidance to affected communities and others.

Good public relations in the general sense also constitute a useful asset for disaster management authorities. This applies especially to the development of public awareness, and thus to the constructive involvement and participation of the public in disaster preparedness, response, and recovery.

## 25. *Mobile Facilities*

The very nature of disaster, particularly its propensity for disrupting existing facilities within affected areas, demands a strong concept of mobility in the application of countermeasures. Therefore, mobile emergency operations centers, mobile welfare facilities, and mobility in such activities as evacuation and survey and assessment need to be fully catered for in organizational terms.

## 26. *Alerting, Activation, and Stand-Down*

Organizational arrangements are necessary for alerting, activation, and stand-down of the disaster management system. These arrangements are usually contained in counter-disaster plans including departmental operational procedures—possibly in stages—as follows:

- Stage 1 - Readiness,
- Stage 2 - Standby,
- Stage 3 - Action, and
- Stage 4 - Stand-down.

See Chapter 12 on plans for details of these stages.

## 27. *Administrative Requirements*

Counter-disaster plans make heavy demands on administrative support. The scale of logistic and administrative backing required for response aspects such as debris clearance, survey and assessment, distribution of relief supplies, and repairs to essential services can be readily appreciated. However, administrative requirements in nondisaster times are equally important, particularly if aspects of preparedness, training, and public awareness are to be maintained at necessary levels of effectiveness. Moreover, day-to-day matters of disaster management must be prioritized if a counter-disaster organization is to function satisfactorily.

## 28. *Emergency Operations Centers*

Virtually all the organizational components outlined above have a direct relationship with the establishment and functions of emergency operations centers. An EOC is the nerve center for dealing with a disaster situation. It brings together the vital aspects of:

- communications,
- information,
- warning,
- situation assessment and monitoring,
- action priorities,
- tasking allocations,
- coordination of effort, and
- media and public information.

In other words, the EOC constitutes the headquarters or focal control point through which the implementation of a counter-disaster plan is directed and coordinated.

In setting up an EOC, the main considerations are:

- adequate communications facilities with a message center;
- arrangements for receiving, collating and assessing information, and for facilitating decision making;
- display facilities (e.g., maps and wall facilities) for presenting an "information picture" of the disaster situation, resources available, tasks being undertaken, tasks needing to be undertaken, etc;
- working space (with desks, seating, etc.) for EOC officials;
- conference/briefing room(s) for briefing ministers and other important persons, and for progress meetings and discussions;
- information room (preferably separate from the main EOC) for briefing media representatives and releasing information to the public;
- rest facilities for staff; and
- other aspects, such as storage space, vehicle access and parking facilities, emergency power supplies, and any other requirements to meet specific circumstance).

Other key points concerning EOCs are:

- It is important that the EOC can be activated at short notice and that designated staff can be readily alerted and mobilized.
- The EOC building must be clearly identifiable.
- The EOC should be periodically checked for serviceability and readiness.
- Alternative EOCs should be earmarked for use if required.

- The tasks and therefore the desk assignments to be carried out in the EOC tend to vary with differing disaster circumstances. This means that designated EOC staff need to come from a wide range of departments in order that action desks can be manned as required by the particular disaster situation.
- One useful way of covering the foregoing manning requirement is to have a roster of disaster services liaison officers (DSLOs) nominated by departments. These DSLOs can then undergo periodic briefing and participation in EOC exercises so that they are ready to undertake their duties any time.

## 29. *Example of EOC Desk Assignments*

The following is an example of the desk assignments and responsibilities which might be applied to an EOC. It is emphasized that this is for purposes of illustration only and that, in reality, desk assignment and responsibilities would need to be formulated according to the needs of the particular situation.

*Controller*
Responsible for the overall direction and coordination of activities undertaken by the EOC. (The Controller would probably be a member of the Central Control Group (CCG) or similar tasking organization, and would implement decisions made by that group in accordance with the policy direction of the NDC).

*Agriculture*
- Collation and assessment of crop damage reports;
- Conservation of produce from damaged crops;
- Immediate rehabilitation measures for agriculture;
- Advice and assistance to producers; and
- Preparations for interim crop production.

*Commerce*
- Liaison with commercial and industrial agencies;
- Location of supplies for emergency government purchase; and
- Monitoring of prices and effects of disaster on local markets.

*Construction*
- Clearance of debris, especially for access and movement;
- Emergency repairs to key facilities (e.g., roads, wharves, and airfields);
- Emergency construction (e.g., temporary shelter, provision of tentage for homeless groups);
- Repairs to government buildings, for essential purposes;

- Repairs to housing;
- Information;
- Collation for information;
- Maintenance of information displays in EOC;
- Dissemination of information, as directed by controller;
- Acquisition of additional information, as required; and
- Media and public relations.

*Medical and Health*

- Management coordination for handling injured and sick persons;
- Preventive health measures;
- Policy deployment of medical teams;
- Policy distribution of medical supplies; and
- Coordination of inspection of food and water supplies.

*Services*

- Restoration of essential services: including power, water, and communications; and
- Direction of special resources to provide services, such as water purification plant.

*Administration*

Accounting, including:

- Relief donations from in-country and overseas;
- Management of national disaster relief fund;
- Government purchases of in-country emergency supplies;
- Clerical procedures;
- Record keeping, for all relevant stores and other items; and
- Handling of immediate claims from public.

*Supplies*

- Monitoring of requirements for emergency supplies (e.g., food, shelter materials, medical supplies);
- Registration of available emergency supplies;
- Arrangement of priorities for distribution of emergency supplies as directed by NDC; and
- Coordination with transport desk of transport and distribution of emergency supplies.

*Transport*

- Coordination of use of available transport facilities with transport requirements.
- Distribution of emergency supplies and other commodities.

*NGO Liaison*

- Coordination of NGO activities.

*International Assistance*

International assistance can be coordinated in several ways. One useful option is for international assistance elements to be channelled through the controller and then dealt with in detail by the appropriate desk; for example, the employment of international medical teams would be dealt with by the controller and the medical and health desk.

## Organizational Structures

30. In this section of the chapter, factors that influence the shape and nature of organizational structures which are necessary to cope with disaster are considered.

### 31. *Basic Considerations*

The following factors are fundamental to determining organizational structures and need to be considered as a first step.

- The physical and geographical nature of the country. For instance, distance alone can be an important factor, as can the multi-island configuration of a country.
- The nature, extent, and severity of the disaster threat, and the broad requirements which these pose in disaster management terms.
- Any formal directions that may have been issued by government concerning disaster management; for instance, legislation or policy statements.
- A broad concept of the scope of activity to be covered by the organizational structure needs to be defined; for instance, whether it is to cover all aspects of the disaster management cycle or only certain parts of it.
- Similarly, a general definition needs to be made of the levels of government structure that will constitute major segments of the structure.
- A general assessment should also be made of the need for and the value and practicability of decentralization. For instance, effective decentralization is a valuable asset if major parts of the organizational structure are put out of action by disaster impact.

### 32. *Identification of Key Government Level*

It is useful to identify, at an early stage in the examination and planning process, what might be termed the key government level—that is, the level within the normal government structure which is the most important for applying disaster management action. This level can vary, depending on individual national circumstances. For instance, in a small country, the key level will invariably be the national level. This is the level from which the main management of normal government affairs is exercised and, therefore, it best fits overall disaster management. However, in very large countries it is preferable to use a state, regional, or provincial level as the key one for disaster management because this tends to be the level from which normal day-to-day government is most actively managed. In fact, because of distance and other factors, it would be impossible to manage disaster response operations from national level.

## 33. *Establishment of Requirements at Various Levels of Government*

Having selected the key level of government, the next step is to establish, in general terms, the main disaster management requirements which should apply at the various levels of government, for instance, at national, state, regional, and local levels.

## 34. *Integration of Organizational and Functional Factors*

It is then possible to integrate:

- Functional factors (disaster management tasks that have to be done) with
- Organizational factors (management framework and facilities needed to carry out the disaster management tasks).

This process leads, in a fairly simple and straightforward way, to the formulation of an appropriate organizational "family tree" as shown in the following diagrams. Figure 1 (also reproduced in Chapter 4) shows the major areas of functional responsibility in a disaster management system. Figure 2 shows a possible structural framework for a large country.

**Figure 1: Disaster Management System Organizational and Functional Diagram**

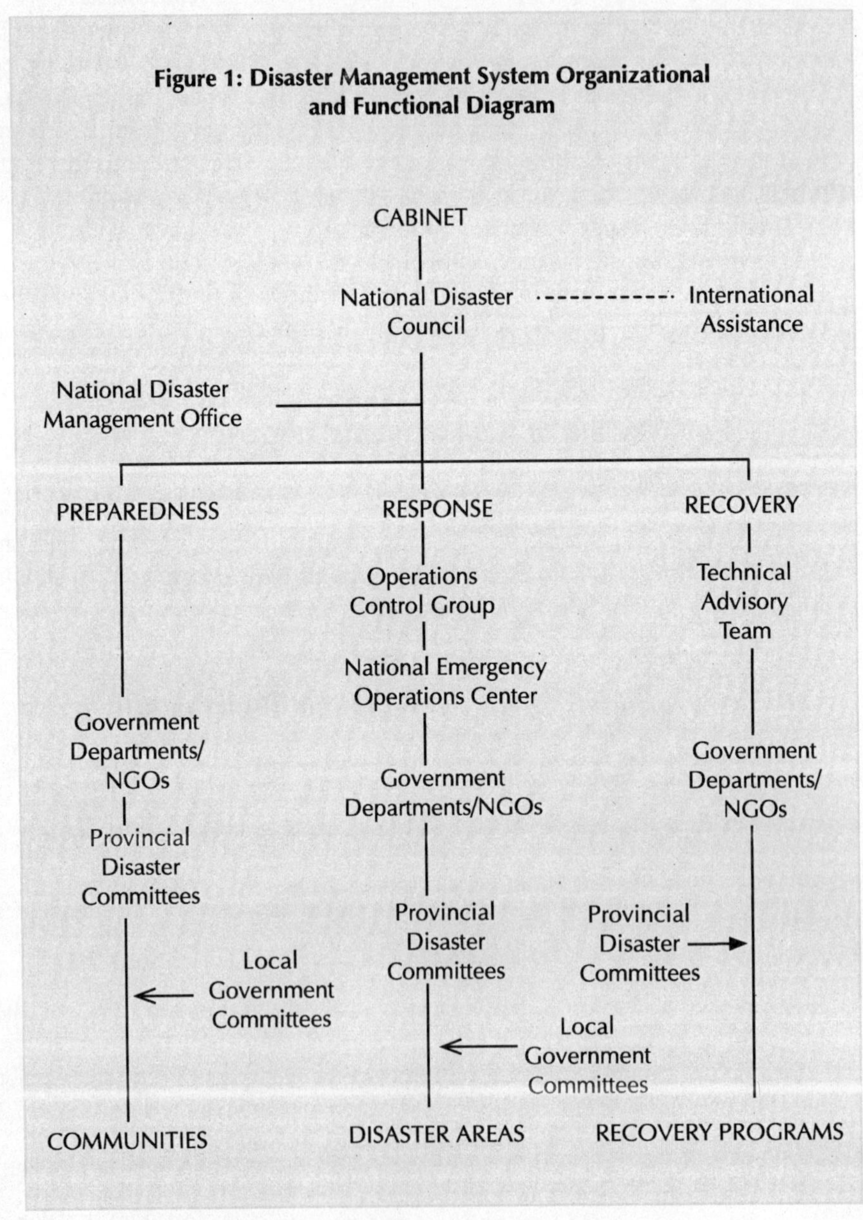


NGO = nongovernment organization

Figure 2: Outline of a Possible Organizational Framework for a Large Country

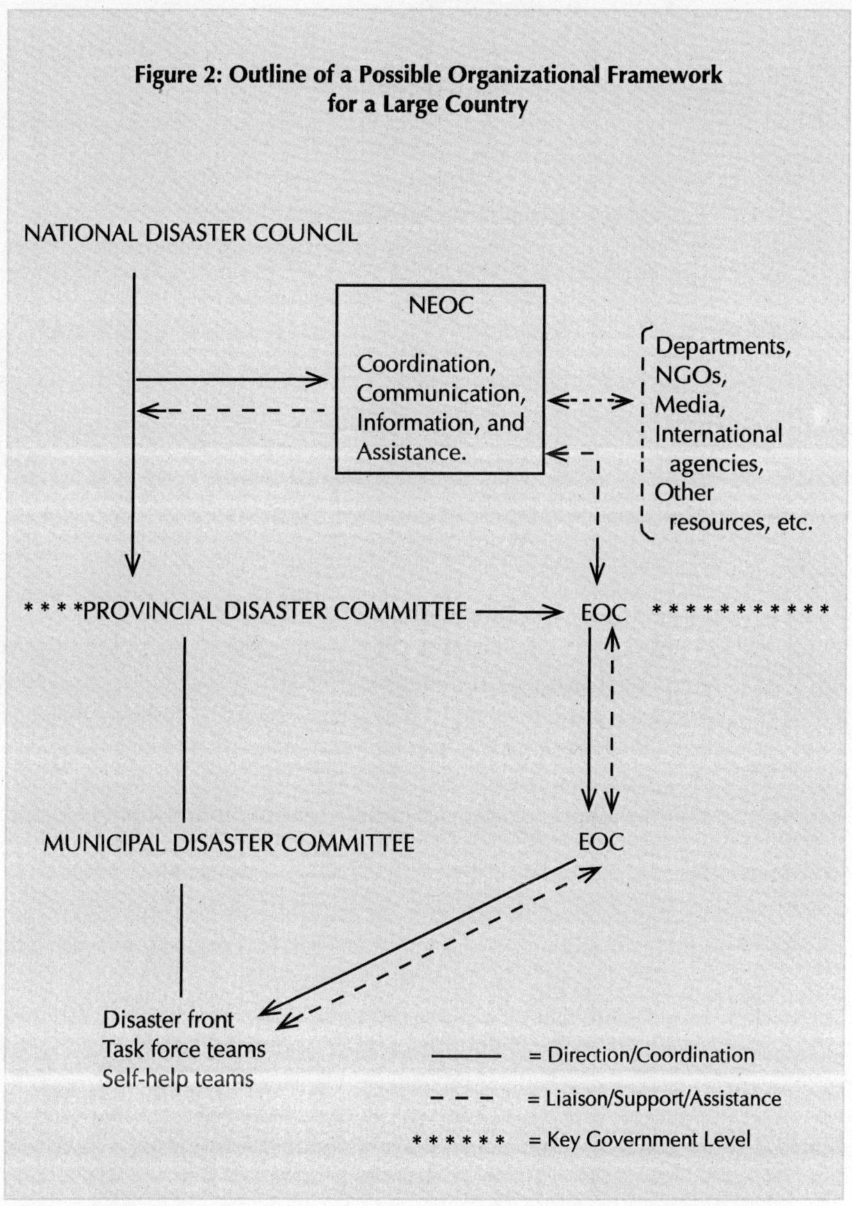


EOC = emergency operations center, NEOC = national emergency operations center, NGO = nongoverment organization.

## 35. *Some Major Responsibilities*

As a further guideline, the responsibilities which attach to some of the major structural components are summarized below. These and associated responsibilities need to be clearly stated in relevant counter-disaster plans.

- *National government responsibility*
  Overall national responsibility for all disaster-related matters rests with Cabinet.

- *Ministerial authority*
  Ministerial authority is vested in the minister [for Home Affairs] who is directly responsible to Cabinet for ensuring that adequate disaster management measures exist at all times.
  In some cases, this ministerial authority is vested in a small ministerial committee of, say, three members.

- *NDC*
  Responsible to the minister (or ministerial committee) for coordinating and directing all disaster-related matters which are concerned with planning, organization, prevention, mitigating, preparedness, response, recovery, training, public awareness, and other appropriate aspects. (The precise responsibilities can vary with circumstances; for instance, the NDC may have limited responsibilities for recovery, which may be dealt with by a separate committee). The council usually consists of a number of heads of departments (e.g., those departments most directly concerned with dealing with disaster such as public works, police, transport, communications, medical and health, foreign affairs), with wide power to coopt other members as necessary.
  NDC can also appoint special bodies, such as task forces. While it is desirable, especially in the interests of timely decision making, that the council should be kept small, it is usually found that in the initial stages of a disaster, it is necessary to call upon coopted members from a fairly wide range of departments and agencies.

- *Operations Control Group (or Central Control Group)*
  Responsible to the NDC for coordination and direction of response operations, including the tasking of resource organizations. The group usually consists of three or four members, for instance:

- assistant commissioner of police (operations),
- chief marine officer, and
- controller of civil aviation.

Apart from their group responsibilities, these officials carry out the duties of controller, NEOC.

- *NDMO*
  Responsible to NDC for carrying out day-to-day disaster management responsibilities, as directed by the chair of the NDC.
  NDMO is usually required to undertake a wide range of responsibilities and it is convenient to include these in an annex to the national disaster plan so that they are available for reference by all concerned.

- *Provincial Disaster Committee*
  Usually, these committees will mirror the NDC in both membership and role. However, where the province constitutes the key government level, the committee would probably be under the chairpersonship of the governor or chief minister of the province.

- *Government Departments*
  Government departments and agencies play key roles in coping with disaster. Therefore, it is important that their responsibilities are clearly laid down. This is best done in department operations procedures.

*Chapter 12*

# Plans

## Purpose

1. The purpose of this chapter is to outline the main considerations which apply to the formulation and use of counter-disaster plans.

2. Certain aspects relating to plans are contained in Chapter 4 and, for ease of reference, these are summarized in paragraph 4 below.

3. Aspects covered in this chapter are:

- Reiteration of the need for counter-disaster plans;
- Some general considerations which apply to planning:
  - the purpose of planning,
  - planning and the disaster management cycle,
  - levels of plans and differing requirements, and
  - the format of plans.
- The planning process;
- Critical areas in planning;
- Aspects that need to be considered during planning; and
- A suggested outline for a main action plan.

Summary of aspects are covered in Chapter 4.

4. Some aspects concerning plans were included in Chapter 4. They are summarized below:

- The need for counter-disaster plans has been well illustrated internationally, the indications being that where plans did not exist, nations suffered more seriously than would otherwise have been the case. Also, an effective basis of planning and the maintenance of relevant plans reflects wide benefits throughout other aspects of disaster management.
- Certain pitfalls can apply to planning, They include;
  - plans becoming outdated and therefore inapplicable;

  - plans being nominalistic and therefore irrelevant; and
  - plans becoming inapplicable due to changes in government organization or similar causes.
- There are several other areas of planning that can prove critical when plans are implemented in response to disaster; these include information management, coordination, and survey and assessment.

## The Need for Counter-Disaster Plans

5. The need for counter-disaster plans, already emphasized in Chapter 4 is reiterated here in the form of two simple quotations.

- The first quotation, already stated in Chapter 4, refers to a disaster which occurred in 1987 and reads as follows: "When we look back on the cyclone and what it did to our country, there is no doubt that hundreds of people are alive here today because we had proper disaster plans."
- The second quotation refers to a disaster which occurred in early 1990. It reads: "A severe cyclone struck the province. UNDRO reports that whilst this cyclone was more severe than the one in 1977, which killed 10,000 people in the same area, casualties are less because of timely warnings and an evacuation campaign."

## Some General Considerations Applicable to Planning

### 6. *The Purpose of Planning*

The purpose of planning is to anticipate future situations and requirements, thus ensuring the application of effective and coordinated countermeasures. It is suggested that this is a useful definition for disaster management officials because it indicates the wide nature of and requirements for counter-disaster planning. In other words, planning should not be confined merely to preparedness for and response to specific disaster events; it should cater, as far as possible, for the whole scope of the disaster management cycle, as outlined in the following paragraph.

### 7. *Planning and the Disaster Management Cycle*

Consideration of planning requirements in relation to the disaster management cycle throws some interesting light on (among other things) the need for flexibility of approach.

*National development*

Most countries nowadays gear national development to a series of time-period plans; for example, 5-year plans. The obvious merit of this kind of system is that it provides considerable flexibility for adjustment to unscheduled or unexpected events; for example, such things as nationwide failure of a vital export crop or unanticipated effects of world financial systems. This flexibility is, of course, also appropriate for adjusting to the effects of disaster. Thus, many nations include disaster aspects in their planning cycles. There is also the point that management of the environment rates highly in modern national considerations. Therefore, since many disaster events are environmentally related, there is a strong case for linking disaster and the environment where national planning is concerned. Consequently, a planning key point is that, where appropriate, disaster should be linked with the environment and included in national development plans.

*Prevention*

Not all countries use the category of prevention on its own, preferring the combined term of prevention/mitigation.

As indicated in Chapter 5, the possible range of preventive measures could be quite wide. At one end of the prevention range, the construction of flood control structures could involve extensive effort and very large amounts of money. For instance, flood barriers built in England over recent years to protect the city of London cost some 700 million pounds (or approximately 1,173 million US dollars). With regard to definitions, it is interesting to note that the system is officially described as being designed to prevent potentially disastrous flooding over more than one third of the capital city area.

At the other end of the prevention range, controlled burning off in forest areas, prior to a high-risk season, to prevent fires from starting, comes closer to mitigation or even preparedness.

Planning for these different contingencies therefore tends to fall into differing categories. For instance:

- A complex and costly flood prevention system, as quoted above, could reasonably be expected to come within the category of national development.
- However, controlled burning off would be more likely to fall within a specific annual disaster management program, which could also be usefully included (for reference purposes), in a disaster preparedness/response plan.

### *Mitigation*

If the term mitigation, or prevention/mitigation is taken as mainly including structural and nonstructural measures designed to reduce the effects of disasters when they occur, it would seem appropriate for such measures to be applied as a series of programs or regulations, rather than as plans. For instance, aspects such as building codes, land-use regulations and safety codes for transport systems would appear to fit more appropriate into a program or regulation category.

However, as with measures of prevention, it would also be reasonable to include appropriate references in disaster preparedness/response plans. For example, the fact that wind-resistant factors had been built into domestic houses would have some bearing on disaster response management decisions relating to possible evacuation or temporary movement to safe havens.

### *Preparedness/response*

The combined categories of preparedness and response tend generally to constitute the most widely used basis for counter-disaster plans, especially those which might be called action plans. This is understandable because so much of response effectiveness obviously depends on good preparedness. Moreover, the use of a plan for these purposes appears well justified by the Oxford Dictionary definition that a plan is a formulated organized method by which things are to be done.

In some cases, the preparedness/response plan is called the (national or state) disaster response plan, as distinct from a separate plan designed to deal with recovery (see comments below concerning recovery).

### *Recovery*

There are various planning options that can be used for recovery. Sometimes a separate plan is utilized so that two main plans exist, a disaster response plan and a disaster recovery plan.

However, some nations prefer to take a more flexible approach and deal with recovery through arrangements which, depending on circumstances, are specific to each disaster event. For instance, recovery matters might be handled by a small ministerial committee.

If arrangements for recovery are left flexible in this way, it is useful to include a small section in the disaster response plan to this effect. There should also be an instruction in the disaster response plan that is aimed at ensuring that, during the response stage, adequate records are kept for post-disaster review and recovery program purposes. In other words, it is essential to interlock response and recovery plans/programs in such a way that both these sections of the disaster management cycle are adequately and effectively processed.

## 8. *Levels of Plans and Differing Requirements*

As indicated in the foregoing paragraph, the different parts of the disaster management cycle generate different planning requirements. In a similar way, levels of plans produce other requirements, as the following examples illustrate.

*National level*
As stated in Chapter 11 on organization, it is not usually practicable to manage disasters directly from a national level in a very large country because of distance and other factors. In this case, the plan (or section of the plan) which applies to the national level is likely to be broadly concerned with coordination, mobilization, and deployment of national resources, requests for international assistance, and so on, rather than detailed management action.

If, however, the country is small and detailed government business can be largely managed from national level, then the national disaster plan is likely to be the main action plan. Therefore, the plan will need a broad scope, ranging from some policy matters down to detailed action requirements.

These aspects of national level planning need careful attention when plans are being initiated, or when they are being reviewed/revised.

*Intermediate level*
In most cases, the plan at intermediate government level will fall into one or two broad categories.

- *If the national plan is the main action plan* (i.e., for a small country), the plan at intermediate level will tend to be more concerned with local implementation measures (e.g., on-ground survey and assessment and detailed relief), rather than major decision making (this latter being the responsibility of national level).

  The intermediate plan will therefore tend to be a smaller projection of the national plan. In such cases, especially in the interests of standardization, it is appropriate for the national government to require that intermediate (say, provincial) plans are formulated to a specified common format. Sometimes a planning guide is issued for this purpose. The actual writing of the intermediate plan then conforms to a set of guidelines and production of the plan is relatively easy and straightforward.

- *If, alternatively, the intermediate level plan constitutes the main action plan*, it will obviously need to be more comprehensive and autonomous than in the circumstances described immediately above. This will particularly apply to aspects such as warning, information management, decision making and control, and coordination capability.

*Community or local level*

At community or local level, plans obviously tend to cover a range of detailed countermeasures. One main requirement of the local plan is, therefore, to coordinate the activities of various existing services (e.g., police, fire brigade, ambulance service, and voluntary organizations). Similarly, the plan needs to facilitate the participation of self-help groups and community members, especially where this encourages the use of traditional knowledge and skill, plus previous disaster-related experience. It is significant that international experience, in both developing and developed countries, underlines the need for and immense value of appropriate plans and procedures at local level.

## The Format of Plans

9. There is no rigid or standard format for counter-disaster plans. The reverse applies such that the format needs to fit the circumstances in which the plan is being made and the requirements which it is designed to meet. However, certain common features tend to apply to most plans. These include:

*Clarity of aim*

The aim needs to be carefully and accurately selected because it determines the whole thrust and scope of the plan. All information, directions, and instructions included in the plan then need to be in line with the aim.

*Realism*

The plan must be realistic such that it relates to an accurate assessment of the disaster threat and considers the scale and capability of available counter-disaster resources.

*Level of the plan*
The plan must also be accurately related to the level with which it is concerned (see paragraph 8).

*Flexibility*
Because disaster circumstances tend to vary and not necessarily follow set patterns, counter-disaster plans need to be flexible. Flexibility is best achieved by planning to cope with the full range of possible disaster threats and ensuring that, within plans, response arrangements can be rapidly adapted to new and changing circumstances. Planned decentralization, where appropriate, is a useful adjunct toward achieving flexibility.

*Coordination*
Since coordination of effort is a key factor in counter-disaster activities, the plan should include the optimum system for direction/coordination. All arrangements covered by the plan should be made with the coordination factor in mind.

*Definition of responsibility*
It is critically important that responsibilities are clearly defined in plans. This reduces, to a minimum, the possibility of misunderstandings, duplications, and omissions in the various activities which the plan covers. It is equally important that responsibilities be clearly defined because it significantly helps in achieving coordination of effort.

*Ease of use*
The plan should be formulated in such a way that it is easy to use. References within the plan should be clear and readily identifiable. Also, the body of the plan needs to be clear and concise as possible, with annexes being used for very detailed information.

*Plan components*
There are a number of options for dividing the plan into sections or components. One way is to have:

- *A main plan or main action plan* that contains the primary parts of the plan, such as the threat, the main requirements for dealing with the threat, resources, organization, direction and coordination, warning, operational implementation of the plan, counter-disaster operations, recovery policy, post-disaster review, etc.

- *Sub-plans* are part of the main plan but which may be required to amplify parts of the main plan that need special consideration (e.g., welfare, evacuation, public information, and so on).
- *Special plans* may be required to deal with special contingencies such as an outbreak of an exotic animal disease, which would require specialist personnel and procedures. Such special plans would normally be designed to work in harmony with the main plan and use the overall counter-disaster organization as necessary.

See also the appendix to this chapter which contains a possible outline for a main action plan.

*Viability*
The plan should include arrangements for periodically checking that it is kept up-to-date and fully viable for the purpose for which it is designed.

*Checkpoints*
It is widely accepted that a counter-disaster plan should be:
written – or it will not be remembered,
simple – or it will not be understood,
disseminated – or it will not be in the hands of those who need it,
tested – or it will not be practical, and
revised – or it will not be up-to-date.

## The Planning Process

10. The process of producing a counter-disaster plan should never be regarded as one in which some specialist or team acts in isolation. The planning process needs to be action-oriented, to involve a wide range of people and organizations, and to produce a result which has the agreement and support of all involved.

11. Some reference points which are applicable to the revision, as well as initial drafting of plans are as follows:

- It is vitally important, throughout the planning process, to keep the aim of the plan clear. After the aim has been initially drafted, it is useful to refer it for confirmation to the minister (or ministerial committee) concerned, together with any major planning guidelines which may apply. This confirmation then provides an official basis on which to proceed for drafting purposes. It is also useful during drafting to refer back occasionally to the minister (or committee) to keep the latter informed of progress and/or to raise any points on which further guidance is necessary.

- It is also important, at an early stage in planning and prior to detailed discussions, to circulate a memorandum to all departments and organizations involved. This memorandum should clearly state the aim of the plan, general outline of the planning program, and anticipated time scale leading up to government approval of the plan.
- It should be remembered, at all times, that planning is a cooperative process. All concerned must be fully consulted, particularly to ensure that mutual agreement is reached on responsibilities designated within the plan. This consultative process is best carried out, from both a practical and psychological viewpoint, by the planners going to see the key individuals concerned, not vice versa.
- After agreeing on responsibilities and undertakings, it is useful if the planners submit to the department or organization concerned a written draft of the proposed input into the plan for confirmation.
- As already stated, there is no rigid format for plans. Therefore, particularly during the consultative process, the planners should develop a good idea of what format is eventually going to be best suited for fitting together the various resource organizations and other components which make up the plan.
- During drafting, it should be remembered that the plan must be easy to use. In this regard, the appropriate use of annexes is essential. Also, sections and paragraphs within the plan should be numbered so that there is no possibility of misreference. In most instances, the plan will be used under difficult disaster circumstances when, perhaps, communications are adversely affected; therefore, everything possible should be done during planning to assist the user.
- Whether the plan should be in a loose-leaf form or otherwise should be considered. The views of experienced disaster management officials sometimes differ significantly on this point. However, from a planning point of view, it should be up to the planners (in consultation with the potential users) to resolve this question satisfactorily.
- How planning progress is to be monitored is also a matter for individual circumstances. One useful method is for the planners, after their initial consultative discussions, to submit a draft outline of the plan to, say, the NDC (acting on behalf of the minister). NDC can then give an overall direction on the general scope and shape of the plan. Subsequently, the planners can draft sections of the plan in detail and arrange periodic progress meetings with the council. This process is useful because it directly involves the council (as

a senior decision-making body) and helps ensure that, for all concerned, redrafting is progressive. This avoids drastic revisions when the draft plan is eventually completed.

- Again depending on circumstances, when the final draft is complete, the minister and/or NDC may deem it appropriate to circulate copies to the departments and agencies concerned for final checking. However, in many cases, this final checking process can be adequately covered if the draft plan is circulated to ministers (and therefore heads of departments) well in advance of consideration by cabinet.
- The planning process and the plan itself should include provisions for legal authorization, thus making the plan a lawful instrument of government. It is recommended that this should be done whether or not disaster legislation exists. The final decision on this point depends, of course, on national circumstances.
- It is worth bearing in mind that planning does not stop once the plan has been finalized and approved. In fact, it is certain that if the plan does not receive constant attention it will rapidly fall out of date. Thus it will be useless, or next to useless when it is next required. In any case the plan, if it is to remain viable, will require ongoing correlated action. For instance, it is necessary to test the plan, or various parts of it, from time to time. Also, facilities such as EOCs require periodic readiness and functional checks. Similarly, important support aspects, such as training and public awareness, need to be kept active. If, in this way, the plan becomes recognised as the focus of disaster management, its viability will be maintained (see also paragraph 20 below for further points concerning viability).

## Critical Areas in Planning

12. As stated in paragraph 4 above, certain pitfalls can apply to planning and the following have already been cited:

- Plans become outdated and therefore ineffective.
- Plans can be nominalistic and therefore lack relevance.
- Plans can become inappropriate due to changes in government organization or similar causes.

13. In addition to this kind of pitfall, there are also various critical areas which need to be counteracted as far as possible. These are outlined subsequently.

## 14. *Crisis Pressure*

The crisis pressure which frequently arises from disaster impact is a potential destroyer of planning effectiveness. Counter-disaster planning usually takes place in slow time, in a calm and structured environment. However, disaster is the antithesis of this. Disaster is usually nasty and dirty, and disruptive and unpleasant. Thus, the crisis pressure generated by disaster impact has all kinds of unwanted effects. For instance, it may cause:

- Loss of, or major disruption to, vital communications;
- Destruction or delayed availability of planned resources (e.g., transport, relief supplies); and
- Disruption to the very system which has been designed to deal with disaster (i.e., the counter-disaster organization).

Very important also is that crisis pressure may have a very adverse effect on people—that is, not only to victims but disaster management officials as well. For instance, cases are known where key officials within disaster warning systems were so badly affected by the onset of disaster impact that they became badly in need of urgent medical attention. Similarly, funding agency officials who had been suddenly implanted into traumatic post-impact conditions which they had never before experienced were, again, so badly affected that it became difficult for them to carry out allotted tasks.

Crisis pressure may be exacerbated by interventions from politicians, which interrupt and disrupt official management. It is not suggested that all problems of crisis pressure can be eliminated or fully counteracted all the time. However, if account is taken of these problems during planning, this can usually go some way toward reducing or offsetting their effects on disaster management capability to handle the particular disaster situation. For example, possible countermeasures might be found in personnel management (i.e., exposure periods to risk) or the timing of response operations immediately following disaster impact.

One difficulty about crisis pressure is that it is very difficult to simulate during exercises or tests. There is no doubt that exercise settings can be made very realistic by the use of gory-looking "casualties," and by producing lots of noise, dirt, and smoke. However, the precise effect of disaster impact on individuals—especially for the first time—is very difficult to predict.

## 15. *Octopus Effect*

Another critical aspect concerns what is sometimes called "octopus effect." This is closely related to crisis pressure and can sometimes be a part of it. However, it warrants separate mention.

It is very easy to assume during planning that the counter-disaster system (whatever it may be) will remain intact and operative and, therefore, capable of implementing planned countermeasures.

However, if the system itself is directly hit, serious problems are likely to arise. This analogy is, of course, that if the octopus gets hit on the head, its tentacles of organization and capability may become paralyzed, albeit only temporarily. For instance, a small country was hit by a very destructive tropical cyclone. The cyclone impacted directly on the capital and inflicted extremely severe damage. This literally paralyzed the government, and its organization and systems. Indeed, for 2 whole days little could be done to implement an organized response.

Fortunately, the country did have an up-to-date plan. Moreover, the plan was based on a system of regional disaster committees. This meant that in the event affected regions outside the capital city area were able to fend for themselves in a reasonably organized way until the total national system could be geared up. This also makes the point that a degree of decentralization (and thus local self-dependence) is useful in planning generally.

## 16. *Information Management*

It is generally accepted that effective information management is of utmost importance if successful response to disaster is to be achieved.

Unfortunately, in many cases, information management proves to be substandard, usually because of inadequacies in planning. The requirements for planning and implementing an information management system are quite clear. They (plus the interrelated systems of communications and warning) are covered in Chapter 11 on organization. Fundamentally, these requirements depend on effective functioning of the disaster information cycle comprising:

- information in,
- assessment or evaluation,
- decision making, and
- dissemination of information and decisions.

All this, as stated above, is quite clear. However, it does call for careful detailed planning because many components, in the form of facilities, systems, and personnel are involved. For instance, if the information management system, as designated in the plan, depends on the temporary conversion of certain office accommodation for EOC purposes, then some important prerequisites are involved. These usually include methodical conversion arrangements, extra display facilities, additional communications, and designated emergency staff. If these prerequisites are not planned in detail beforehand, breakdowns of various kinds are certain to occur. This aspect, in fact, repeatedly occurs as a shortcoming in various disaster analyses. The point should be carefully noted by disaster management planners.

## 17. *Survey and Assessment*

Survey and assessment requirements also call for careful provisions within plans. Again, the requirements are clear, but they are detailed and they do require precise arrangements. They also involve significant logistic, administrative, and training aspects. One key characteristic of survey and assessment—from a planning viewpoint—is that several different specializations tend to be involved (e.g., medical and health, essential services, and agriculture) and therefore effective coordination is required in:

- survey and assessment operations,
- the information that is acquired, and
- the resultant response and recovery action.

Some of the planning issues that need to be covered for survey and assessment are:

- Definition of the data likely to be required following disaster impact (e.g., casualties, damage, remaining resources/facilities);
- Means available for collecting and recording information (e.g., air survey, ground survey);
- Provision and ready availability of survey and reporting proformas;
- Direction and management of survey and assessment activities; and
- Training and periodic briefing of survey teams.

Certain publications are available on survey and assessment and disaster management planners would be advised to study some of these. However, provide a basis of information, Annex C of this handbook contains a general overview.

## 18. *Definition of Responsibilities*

The importance of clear definition of responsibilities has already been stressed as a key planning guideline. Its critical nature is reiterated here. It calls for meticulous and often painstaking planning and there is no short cut on the consultative process if all responsibilities are to be satisfactorily defined and allocated. As indicated in Chapter 8 on counter-disaster resources, the planning "legwork" involved in satisfactorily achieving this goal can be considerable.

## 19. *Coordination*

As with allocation of responsibilities, coordination aspects can be critical when plans are implemented, unless prior detailed planning has been very carefully and accurately carried out. Some of the aspects which produce problems or failures in coordination are:

- Unclear coordination responsibilities;
- Inadequate provision, maintenance, and preparedness of facilities, mainly EOCs;
- Inadequate activation arrangements;
- Loss of or severe disruption to vital communications links and inadequate planning for reserve links;
- Poor liaison throughout the counter-disaster system;
- Discrepancies in information gathering and information processing, leading to inaccurate situation assessment and decision making;
- Inadequate direction of resource organization by EOCs;
- Failure of resource organizations to carry out allotted tasks accurately and effectively;
- Loss of operational coherency by EOCs and/or resource organizations; and
- Loss or non-availability of key leaders and inadequate planning arrangements for substitutes.

It is important to ensure during planning that optimum arrangements are established for direction/coordination. However, in practice, effectiveness in coordination can vary markedly. This can happen for a variety of reasons, such as:

- Loss of experienced officials for various reasons;
- Lack of practice in exercises and simulations;
- Limited opportunities for participation in actual operations; and

- Lack of appreciation by senior disaster management officials of the key importance of coordination.

Therefore, not only must the plan contain the best possible arrangements to facilitate and ensure coordination of effort, but continuous attention must be given to it. It does, in fact, provide the underlying link through the various factors that help maintain planning viability.

## 20. *Viability of Plans*

Once the plan has been finalized and approved, it needs continuous attention, especially to maintain its viability. Maintaining the viability of plans should be one primary responsibility of the office or section within government which is charged with handling day-to-day disaster management affairs (e.g., an NDMO). To carry out this responsibility, it is helpful to formulate and work to a definite program or series of programs. Such programs, which obviously need the authority and support of the NDMO (or the equivalent), should include aspects such as:

- *Exercises and tests*
  These can comprise major exercises aimed to test and train the overall counter-disaster organization; or they can be restricted to certain aspects, such as activation. It is, of course, necessary to use the results of exercises and tests to improve the plan, wherever this is possible.

- *Functional and readiness checks*
  These need to be carried out periodically to ensure that key facilities and systems are serviceable and ready for use (e.g., EOCs, emergency shelters, welfare centers).

- *Training activities*
  All training activities, in addition to exercises and tests, should be based on the plan. This develops and maintains awareness of the plan and its key role in disaster management generally.

- *Post-disasterr eview*
  As stated in paragraph 9, the plan should contain arrangements for ensuring that it is periodically checked and kept up-to-date. In this regard, post-disaster review provides a valuable means of checking the plan and amending it as required.

- *Use of regulatory measures*
  In some circumstances, regulatory measures can be applied to help ensure that plans are maintained and kept under review. These include:
  - ensuring that government departments maintain their own action plans, using a standard format;
  - directing departments to report their state of readiness to the NDC (or its equivalent); and
  - withholding annual budget allocations to departments until their state of readiness has been certified.

- *International assistance liaison*
  The plan should contain the arrangements that apply to international assistance and the liaison channels which are to be used. If, as should be done, ongoing dialogue in various forms is maintained with international assistance agencies, according to the plan, this means that the latter is being partially exercised in non-disaster times.

- *Public awareness activities*
  Some public awareness activities offer a good opportunity for maintaining the viability of the plan. For instance, a National Disaster Preparedness Day can help raise levels of awareness for both official agencies and the community. The day needs to be widely publicized and set at themost appropriate time such as before the flood or cyclone season. On this day, all government officials having responsibilities under the plan would check all aspects of planning, preparedness, and response relevant to those responsibilities. NGOs would be asked to do likewise. NDC would meet to consider the general preparedness situation if a disaster struck on this day. EOCs would also be given a test activation. Action could also be initiated at local government level and in schools; for instance, a simulated evacuation of an area or a school building. These are, of course, only simple examples. The point is that if officials and the community can be involved in this type of activity, then automatically various aspects of the plan's viability are being tried out.

- *Publicity on preparedness*
  If the disaster threat or part of it is seasonal, this provides a good opportunity to give realistic publicity to preparedness measures contained in the plan. Suitable media coverage for this therefore assists in maintaining viability.

- *Education in schools*
  Many countries have introduced appropriate disaster education into school programs. This, again, needs to be based on relevant plans and therefore assists community awareness.

- *The plan itself*
  The plan itself can also contribute to viability. If it is attractively presented, clear and easy to read, people will refer to it. On the contrary, if the plan is a muddled, tired-looking document, overwhelmed by masses of statistics, no one will bother to look at it, let alone read it.

One of the most effective means of maintaining the viability of plan is, of course, regular operational use. Therefore, (and unfortunately in a sense) countries and areas which face a high level of disaster threat are likely to have less problems in maintaining viability than those in which the threat is low level and infrequent.

### 21. *Planning Responsibility*

Obviously, the responsibility carried by disaster management planners is onerous. If the planners get the plan wrong, then the repercussions can be very severe and widespread, possibly involving the loss of many lives.

On the other hand, accurate and meticulous planning not only produces an effective plan, it also provides the focus for successful disaster management overall.

## Aspects for Consideration During Planning

22. The purpose of this section of the chapter is to provide a reference for the formulation of counter-disaster plans by listing major aspects that need to be considered.

23. The planning process usually involves consideration of a wide range of disaster-related matters to decide what is eventually included in the plan. However, not all aspects will be relevant to all levels of plans. Neither will all aspects assume equal importance in different plans. As a simple example, details of shelters (including their readiness and management) will be very important at community level but will have little or no relevance to a national level plan. The relevance of various aspects therefore needs to be subjected to careful analysis and sifting by planners.

24. Most of the aspects mentioned in paragraphs 25–49 below will need to be considered during the process of formulating a counter-disaster plan. In practice, this consideration is not chronological; it requires constant cross-reference and adjustment between the various aspects. Also, there is no set point in the planning process at which it is best to draw up an outline of the plan. Generally speaking, the sooner a reasonably firm outline is available, the easier it is for the planners to begin putting together the component parts. However, in some cases, the disaster management authority (the minister and/ or the National Disaster Council) may require a preliminary outline plan before authorizing planning to go ahead.

## 25. *Policy Direction*

All relevant resources of policy need to be examined. These might include:

- National development plans,
- National disaster management policy statement, and
- Specific directions concerning the formulation of counter-disaster plans.

## 26. *Legislation*

This might include:

- Special disaster legislation,
- Emergency powers, and
- Any other relevant regulations.

## 27. *The Disaster Threat*

This obviously needs to be carefully analyzed and to cover all possible eventualities, taking into account action that has been or is being taken with regard to hazard identification, vulnerability assessment, and risk evaluation.

## 28. *Disaster Scenarios*

It is often helpful to consider specific disaster scenarios that might emanate from the disaster threat. This enables planners to obtain a focus on particular eventualities, in terms of examining warning, activation, mobilization of resources, initial response operations, and the possible scale of total response. Some planners advocate the inclusion of disaster scenarios in the plan itself but others argue that this may encourage a set, inflexible approach to use of the plan. Choice in this matter depends largely on individual circumstances.

## 29. *Likely Problems Arising from the Threat*

These problems which would, of course, come under scrutiny in consideration of possible disaster scenarios can be defined fairly easily, (see also Chapter 2). The important thing, in relation to the plan, is that the problems should be realistically related to the actual components of the threat. This may, in fact, give some indication of which priorities need to be given to countermeasures but this very much depends on local circumstances.

## 30. *Scope of Disaster Management Effort*

Governments sometimes find it necessary, for financial and other reasons, to place limitations on the scope of disaster management activity. Where this applies, plans are likely to be restricted accordingly; if so, any applicable restrictions should be indicated in the plan.

## 31. *Aim of the Plan*

The aim of the plan must be accurately defined. It is also important to define the aim clearly, preferably in a single sentence. For example:

> The aim of this plan is to define the action to be taken to deal with disaster in Exland.

The simple and clear definition does not preclude qualifications being made concerning the circumstances in which the plan might be invoked. For instance, a qualifying statement might emphasize that the plan is not designed for day-to-day incidents which are within the capacity of the normal emergency services; or that the plan would only be invoked after a decision by NDC.

## 32. *Definitions*

Definitions used in or applicable to the plan need to be considered and resolved, so that an appropriate reference can be included in the plan itself.

## 33. *Main Requirements for Dealing with Disaster*

From consideration of:

- policy directions,
- the disaster threat,
- possible disaster scenarios,
- likely problems arising from the threat, and
- the scope of disaster management effort,

it should be possible to define realistically the main requirements for dealing with disaster such as warning, rescue, and survey and assessment (see also Chapter 4 on Major Requirements for Coping with Disaster).

## 34. *Counter-Disaster Resources*

Having defined, as accurately as possible, the likely requirements for dealing with disaster, it is logical to examine the resources available to implement these requirements. This is a major part of the planning process and needs to cover all possible availabilities of government, nongovernment and international assistance resources. No possibility should be overlooked.

A case is known where a light aircraft, belonging to a church body and kept on a remote private airstrip (plus the church's radio communications network) proved to be a vital counter-disaster asset. Annexes to the plan need to cover a list of all resources applicable to the plan and allocation of roles and responsibilities to resource organizations.

## 35. *Organization*

Consideration of the type of organization needed for any particular plan constitutes another major part of the planning process. In this regard, two important requirements usually need to be carefully examined:

- First, the organization should fit simply and effectively into the existing government organizational structure.
- Second, it should be capable of changing as rapidly as possible from a normal day-to-day role into an operational one.

Responsibilities also need to be defined for:

- The responsible minister (or ministerial committee),
- NDC,

- OCG,
- Special advisory teams (if applicable),
- NDMO,
- Relative arrangements at other planning levels, and
- International assistance arrangements.

These responsibilities should cover both nondisaster and disaster conditions.

See also Chapter 4 on major requirements for coping with disaster. Scrutiny of relevant plans from other countries may also provide some useful guidelines toward definition of the organizational system.

## 36. *Preparedness Measures*

Preparedness measures relevant to the functioning of the plan need to be considered and designated as follows:

- Types of preparedness measures,
- Levels at which the measures apply,
- Responsibilities for implementing the measures, and
- Processes by which the measures are kept updated and revised.

Special preparedness aspects—where they apply—such as emergency evacuation procedures and shelter arrangements may be best included as annexures or subplans.

## 37. *Communications*

As stressed in Chapter 11 on organization, and as strongly indicated in other parts of this handbook, communications constitutes a vital disaster management component. Nowhere is this more evident than in a counter-disaster plan. Without effective communications, it is impossible to implement a plan. The planning process therefore needs to cover communications aspects very carefully and thoroughly with specialist advice, as required. Communications resources that are best included in a self-contained annexes to the plan need to include:

- Government services (e.g., police networks, national telecommunications);
- International services (e.g., aeronautical and meteorological networks); and
- Private networks (as appropriate).

The aim in planing should be to provide disaster management officials with the best possible communications options, plus maximum stand-by capability in case of failures.

## 38. *Operational Direction and Coordination*

Arrangements for the direction and coordination of operations need to cover aspects such as:

- Responsibilities,
- EOC facilities and their locations, and
- Staffing arrangements.

## 39. *Warning*

The warning information contained in the plan obviously needs to be clear and unambiguous. It should include:

- Types of warning relevant to various forms of disaster threat;
- Sources of warning;
- Distribution of warning information to key officials and agencies;
- Dissemination of warning to the public; and
- Notification of "All Clear" if, for any reason, the threat does not materialize.

## 40. *Implementation of the Plan*

Planners should consider the inclusion in the plan of a system of implementation, probably in stages such as:

- Stage 1 Alert (or Readiness),
- Stage 2 Stand-by, and
- Stage 3 Action.

The advantages of a system such as this are:

- It is a convenient way of informing disaster officials and the public of the progress of the disaster threat and the action being taken.
- It helps ensure a graduated and methodical response to the disaster situation.
- It also helps to avoid overmobilization of resources, if the disaster threat does not develop.

Obviously, in no-warning situations, Stages 1 and 2 become non-applicable and this needs to be explained in the plan.

## 41. *Operations*

This section needs to provide a form of operational checklist for the user of the plan. It therefore needs to be comprehensive but as succinctly as possible. It should include:

- Precautionary measures on receipt of warning;
- Activation of EOCs;
- Direction and coordination of operations (operational additions, as required, to paragraph 38 above);
- Operational requirements in the stricken area, for example:
  Survey and assessment,
  Rescue,
  Treatment and care of casualties,
  Clearance and access,
  Communications,
  Evacuations,
  Shelter,
  Food,
  Water and power supplies,
  Temporary subsistence supplies,
  Health and sanitation,
  Security,
  Public information,
  Construction requirements,
  Disaster welfare inquiry, and
  Other requirements specific to the situation.

Also included in this section should be a reminder that certain operational action will probably need to be taken specifically from national level; for instance, requests for special aerial reconnaissance by overseas agencies, operational assistance teams, etc.

The section also needs to include a reminder that the emergency phase of operations should be kept reasonably short. In practice, this emergency phase tends to be about 2–3 weeks, after which the normal systems and processes of government should progressively apply.

## 42. *Recovery*

This section of the plan needs to state briefly the policy for post-disaster recovery and the main measures usually involved. In most cases, recovery arrangements—since they can be very protracted—are made the subject of a separate recovery plan or the responsibility of a ministerial committee. However, inclusion in this plan of certain key information concerning recovery provides a form of disaster management link.

## 43. *Post-Disaster Review*

Post-disaster review is very important to ensure that future disaster management arrangements are developed satisfactorily.

This part of the plan therefore needs to cover:

- Responsibility for post-disaster review;
- Subject areas to be covered;
- Debriefing of individual departments and agencies;
- Review of the plan and the counter-disaster organization, and amend as necessary.

## 44. *Support Measures*

It is advisable to include in the plan a section covering the responsibilities for support measures, such as:

- Training and
- Public awareness.

Also, if required, outline programs for these aspects can be included as annexes.

## 45. *Relationship with Other Plans*

During the development of the main plan, the need for initiating associated plans and standard operating procedures will almost certainly emerge. The relationship with these additional plans and procedures needs to be covered in a section of the main plan.

## 46. *Distribution List*

The distribution list for the plan can be positioned in the front pages of the plan and/or as an annex.

## 47. *Contents*

A clear list of the contents of the plan is important for ease of reference. The contents pages should be positioned at the front of the plan.

## 48. *Authorization and Date of Issue*

These need to be associated with the contents.

## 49 *Map References*

These are best positioned following the contents and authorization.

50. A suggested outline for a main action plan, based on the foregoing guidelines, is shown as an appendix. The outline provides a reasonable basis for the plan at national level for a small country; or at intermediate levels for a large country. It is recognized, of course, that plans based on this suggested outline would require additions and/or deletions to suit specific circumstances.

# Annex to Chapter 12

## SUGGESTED OUTLINE FOR A MAIN ACTION PLAN

CONTENTS

AUTHORIZATION

MAP REFERENCES

INTRODUCTION

The Disaster Threat
National Policy
General Concepts for Disaster Action

AIM OF THE PLAN

DEFINITIONS

RELATIONSHIP WITH OTHER PLANS

MAIN REQUIREMENTS FOR DEALING WITH DISASTERS

EMERGENCY POWERS

Disaster Legislation
Other Legislation

COUNTER-DISASTER RESOURCES

ORGANIZATION

Organizational Structure

- Minister Responsible,
- National Disaster Council,
- Central Operations Group,
- National Disaster Management Office,
- Provincial or Regional Disaster Committees,
- International Assistance Liaison, and
- Nongovernment Organizations,

Organization for Counter-Disaster Operations
Allocation of Roles and Responsibilities

International Assistance Arrangements
Coordination of Planning and Organizational Measures

PREPAREDNESS MEASURES

General
National Level
Provincial Level
Community Level
Training and Public Awareness Programs

COMMUNICATIONS

OPERATIONAL DIRECTION AND COORDINATION

Responsibility for Operational Direction and Coordination
National Emergency Operations Center
Provincial or Regional Emergency Centers

WARNING ARRANGEMENTS

General
Agencies Originating Warning
Transmission of Warning
Dissemination and Public Broadcast of Warning
Notification of All Clear

OPERATIONAL IMPLEMENTATION OF PLAN

Stages of Implementation

COUNTER-DISASTER (OR RESPONSE) OPERATIONS

Precautionary Measures
Activation of Emergency Operations Centers
Direction and Coordination of Operations
Information Requirements
Operational Requirements in Stricken Areas
Operational Action – National Level
Operational Action – Provincial, Regional, and Local Levels
Period of Disaster Operations (Emergency Phase)

RECOVERY

Statement of Policy for Recovery
Responsibility for Recovery Programs
Cross-Reference to Recovery Plan (if applicable)

POST-DISASTER REVIEW

Responsibility
Debriefing
Review of Plans and Organizations

SUPPORT MEASURES

Training
Public Awareness

VIABILITY OF THE PLAN

Responsibility for Ensuring Viability

ANNEXES

Distribution List
List of Definitions
List of Resources
Functional Diagram of Organization
Allocation of Roles and Responsibilities to Resource Organizations
Guidelines for International Assistance Arrangements
Communications
Detailed Information on Warning
Precautionary Measures on Receipt of Warning
Guidelines on Training
Guidelines on Public Awareness
Format for Departmental Standard Operational Procedures
Duties and Responsibilities of National Disaster Management Office

*Chapter 13*

# Use of Resources

## Purpose

1. The purpose of this chapter is to consider major factors that apply to use of counter-disaster resources. Certain aspects relating to resources are also contained in Chapter 4 on major requirements for coping with disaster and in Chapter 8 on counter-disaster resources. These are summarized in paragraphs 3 and 4 below.

2. Aspects covered in this chapter are:

- the need for optimum use of resources;
- functional requirements of resource organizations;
- management of resources for optimum effectiveness;
- activation of resources;
- graduated response; and
- commodity management.

## Summary of Aspects Covered in Chapters 4 and 8

3. Some aspects concerning use of resources are contained in Chapter 4. They are summarized below:

- There are many difficulties in achieving optimum use of resources. Most of the prevalent ones are listed.
- An important management proviso is to correctly use of management systems within individual resource organizations.
- Effective resource management depends largely on:
  - a capable EOC system;
  - a good information picture;
  - effective communication between the direction/coordination authority and individual resource organizations; and
  - sensible commitment of resource organizations to operational tasks.

4. Chapter 8 covers the following aspects:

- Types of counter-disaster resources.
- Evaluation of resources.
- Examples of roles and responsibilities.

## The Need For Optimum Use of Resources

5. One hard fact of life that emerges from analyses of disasters is that available resources are seldom adequate to cope with requirements. Moreover, the pattern of shortfall or discrepancy is subject to considerable variation, as the following examples indicate:

- Disaster can make very heavy demands on some resources but not necessarily on others. For instance, response to an earthquake situation places high priority on heavy earth-moving equipment, search and rescue, and special medical skills to the extent that these resources are nearly always overburdened.
- In some disaster situations, adequacy of one resource capability may be reduced or offset by limitations in other resources. For instance, in a situation where there are heavy requirements for emergency feeding, food resources may be adequate but distribution capability is not.
- Shortfalls in resources may sometimes be self-imposed. For example, if planning has been inadequate, the full range of communications resources may not be available when urgently required.

6. In addition to the foregoing considerations, it is necessary to bear in mind that:

- Under most circumstances, the possible scope of disaster management activity is wide and varied; therefore, the participation of many organizations, facilities, and personnel is required. This can pose numerous problems in management and use.
- The impact of major disaster imposes severe demands across a wide part, it not all of the national structure.
- Recovery from disaster places a considerable overload on many national resources, often over a prolonged period.

7. There is nothing new, therefore, in situations where individual countries around the world may find themselves pinched for adequate resource capability.

Moreover, it is widely accepted internationally that it is neither practicable nor cost-effective to have special services—as distinct from standard emergency services—standing by just in case a disaster occurs. The alternative, therefore, must be to use existing capabilities. However, when the requirements of preparedness for response to and recovery from disaster are considered and assessed, it becomes abundantly clear that these capabilities have to be optimized. In other words, there is little point in expecting disaster-relevant resources to be effective in their normal role, plus a disaster role, without enhancing their performance to a required level.

8. However, the nature of most counter-disaster resource organizations is such that it is not easy for them to achieve and maintain peak dual-role performance by their own means. For example, most government departments (which constitute primary counter-disaster resources) are rigidly tasked and staffed. Moreover, they tend to be under the continuing pressure of establishment cuts and other forms of economies. Generally speaking, therefore, these kinds of resources will not achieve high levels of effectiveness in their disaster roles if they are left entirely to their own devices. This, obviously, is where disaster management assistance comes in.

9. Given that disaster management assistance has a viable role here, certain other issues need to be borne in mind. These issues relate to the fact that resource organizations tend to vary in regard to compatibility between their normal role and their disaster role. For instance:

- Some organizations are closely oriented toward disaster tasks. As an example, a public works department has the equipment for and day-to-day experience in clearing debris and cutting access roads, which is a primary disaster task. Similarly, much of a telecommunications agency's daily task is concerned with maintaining communications systems; in times of disaster that task may increase considerably but, in essence, it does not change.
- Other organizations, however, may have to make significant adjustments when extending into their disaster roles. A community services department, which is heavily desk-oriented normally may have to switch a proportion of its personnel into active field roles for emergency feeding and welfare center purposes.
- Some resource organizations may have difficulty in extending satisfactorily into their disaster role because of the need to keep their primary role functioning at near-normal level.
- With NGOs, the assumption of disaster roles is often easier than is the case with government organizations. Most NGOs are normally concerned with various aspects of community care and assistance.

They are, therefore, well suited to carry out similar disaster roles and, indeed, in most cases would expect to do so. Moreover, because of their charters, the majority of such organizations would have difficulties in changing roles. From a disaster management viewpoint, this is a mixed blessing because it could, under many circumstances, provide an excess capability in one general role area while, at the same time, precluding the switch of excess resources into other roles.

The above illustrations indicate that if disaster management authorities assume responsibility for ensuring that all resource organizations can convert readily into the disaster mode, they must be aware that complications, such as those above, are involved. Moreover, the illustrations suggest that conversion cannot be standardized and that therefore alternative and flexible methods need to be devised.

## Functional Requirements of Resource Organizations

10. Before examining these possible methods, however, it is advisable to look further at some of the key requirements that affect the functions of resource organizations. These include:

- *Clear understanding of the disaster role*
  It is obviously necessary that organizations should have an absolutely clear understanding of the disaster role(s) required to be fulfilled, especially the additional commitments involved.
- *Organizational scope and limitation*
  The scope and limitation of organizations in relation to their disaster tasks also needs to be examined; for instance, to identify any sections and/or personnel who may not be required to participate in the allotted disaster role. These sections/personnel may be valuable for earmarking as reserve capability of some kind.
- *Role capability*
  The capability of organizations to fulfil their disaster roles must also be examined, monitored, and confirmed; also their ability to sustain the role under disaster conditions.
- *Timely role conversion*
  Organizations must be able to convert to their disaster roles in time to meet tasking requirements; alternatively, there must be adjustment to those requirements.
- *Suspension of normal role elements*
  If some elements of some organizational roles can be temporarily suspended or left in abeyance, these should be identified as clearly

as possible, thus releasing the relevant resources for disaster purposes.

- *Effect of disaster commitments on own resources*
  A factor which often concerns and constrains organizations in relation to disaster roles is the various forms of loss which they may suffer. This can happen in the form of using up fuel, transport capability, supplies and budget allocations. The result is that, after the disaster, an organization's own role capability is depleted or downgraded. Moreover, in conditions of post-disaster stringency, the losses may take a significant time to recoup. This point needs to be resolved, as far as possible, especially to ensure that organizations are not inhibited from carrying out their disaster operations.
- *Exposure to risk*
  Sometimes disaster operations may expose the personnel and equipment of resource organizations to risk or loss. This may not be acceptable to resource organizations, resulting in reluctance to carry out certain tasks. This point, if it is likely to be relevant, needs to be satisfactorily settled during the planning process.
- *Legal Implications*
  As with exposure to risk, disaster operations may sometimes involve disaster workers and/or organizations in legal implications. For instance, damage may be caused to private property during these operations. Resource organizations need to be assured that they are safeguarded against this kind of eventuality. Disaster legislation can usually be framed to cover this kind of possibility.

11. The effect which a disaster commitment might have on a resource organization can be illustrated as shown below.

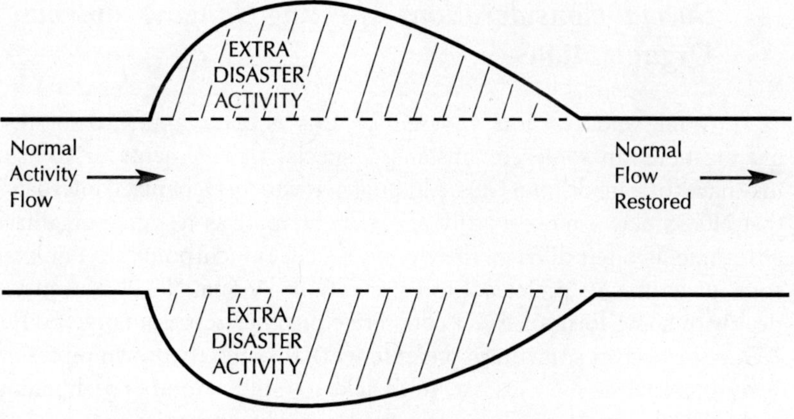

This onion or balloon effect, if it occurs, obviously needs to be controlled and kept to a minimum, otherwise the organization may choke and become useless in its disaster role. Possible ways of dealing with this problem are suggested in paragraph 15 below.

## Management of Resources for Optimum Effectiveness

12. Probably the biggest problem concerning the key requirements in paragraph 10 above is ensuring that they are fully recognized and accepted by both resource organizations and disaster management authorities. Once the requirements are thus clarified, there is no reason why they cannot be resolved by good disaster management processes, as suggested in paragraphs 13–18 below.

### 13. *Rationalization of Roles*

The overall management requirement is obviously to achieve the best possible rationalization of the major aspects that affect both normal roles and disaster roles. It is suggested that this is, to a large extent, an ongoing disaster management task. It may involve a substantial monitoring component, especially if variations and changes are likely to occur in either normal roles or disaster roles. If standard operating procedures (see paragraph 16 below) are maintained, these provide a useful medium for monitoring and checking. Moreover, if such standard operating procedures are officially authorized in disaster legislation and/or counter-disaster plans, they provide a means of formal backing for the disaster management authority and, indeed, for resource organization also.

### 14. *Special Considerations Affecting Nongovernment Organizations*

While most factors covered by this chapter apply to all resource organizations, in some circumstances, special arrangements or adjustments may have to be made for NGOs. Although it is widely recognized internationally that NGOs play a most worthy and valuable part as resource organizations, governments often differ in their views on NGO incorporation. For example, some governments prefer not to liaise with individual NGOs but, instead, to deal with some form of NGO committee or other representative body. Also, NGOs themselves sometimes prefer to work through their own representative body. In such cases, NGOs may undertake to fulfill a number of different roles and let their committee allocate tasks to individual organizations.

From a disaster management viewpoint, the important point here is not so much who undertakes tasks but that these tasks should be fully covered and carried out with optimum effectiveness. Thus, whatever system is in place for NGO participation, there still needs to be a workable arrangement between the disaster management authorities and NGOs themselves for ensuring functional efficiency. It is also important to bear in mind that arrangements for NGO participation need to cover nondisaster times as well as periods of emergency operations.

## 15. *Onion Effect*

Some possible ways of dealing with the onion effect (see paragraph 11 above) may be:

- The shedding of some normal role activity as soon as the need to convert into the disaster role becomes evident.
- Implementing special planning measures to absorb or offset the extra load, particularly during the peak load.
- The training and employment of extra staff; that is, the use of some form of reserve or auxiliary component. For instance, some countries call up and use police auxiliaries during disaster periods.
- The holding of some special stand-by facilities or capability (e.g. equipment, stores) if these are especially applicable to. the organization's disaster role.
- The use of emergency operating procedures which reduce the workload.
- The acceptance of the extra load and the use of all facilities and staff on a crash-program basis, in the hope that the crisis may be shortlived. This is not a recommended management risk and it can be dangerous if the crisis period is prolonged and/or a secondary disaster impact follows closely upon the original one.

## 16. *Standard Operating Procedures*

There is no doubt that if all resource organizations work to standard operating procedures, under relevant counter-disaster plans, this constitutes one of the most effective means of ensuring their optimum utilisation. A possible format for such operating procedures is given in the Appendix to this chapter.

## 17. *Training*

All forms of training which are concerned with the functions of resource organizations, under relevant disaster plans, obviously constitute a valuable means of ensuring optimum use.

18. ***General***

Items such as those included in paragraphs 13–17 may, in themselves, be almost marginal in optimizing the effectiveness of resources. Nonetheless, it is these marginal contributions, which when combined, ultimately ensure such effectiveness. Therefore, consideration of resource capability in this detailed way can produce real and worthwhile benefits.

## Activation of Resources

19. One area concerning the use of resources that may not always receive adequate attention in counter-disaster planning, organization, and training is that of activation. The activities concerned are primarily related to response, though they could apply to certain aspects of preparedness; for example, mobilizing resources to build emergency levee banks in advance of a possibly developing flood situation. In any case, a number of factors may affect the way in which activation is implemented. These factors are best considered against what might be called a best-case situation.

20. ***Best-Case Situation***

In this situation, the following conditions would generally apply:

- Adequate plans would be in place;
- Resource organizations would work to standard operating procedures;
- Plans and standard operating procedures would be known, understood, and regularly practiced by all resource organizations and their individual members;
- There would be an effective warning system, operating through an established EOC system;
- There would be effective communications for operating an EOC system and its resource organizations; and
- Plans and standard operating procedures would contain a graduated implementation system, along the lines of:

*Stage 1 – Readiness (or Alert)*

This stage comes into effect when some form of information is received which indicates that the provisions of the counter-disaster plan may have to be invoked. The national disaster council (or other appropriate body) will declare this stage in force and, if considered necessary, will authorize a public broadcast to this effect. NDC will consult, as appropriate, with other committees (e.g., provincial) involved and will decide what action is to be taken at this stage. NDC will then initiate such measures under the plan as necessary. In some circumstances, when there is no warning, this stage does not apply.

*Stage 2 – Stand-by*
This stage comes into effect when it is established that a threat exists to all or part of the country. It requires that all relevant organizations and personnel are placed on stand-by, so that they are ready to begin operations or action under the plan immediately when they are called upon. NDC, in consultation with other committees (as appropriate) will decide when this stage becomes operative and will announce it accordingly. As with Stage 1, this stage may become inapplicable in no-warning situations.

*Stage 3 – Action*
This stage comes into effect when it appears certain that a disaster is imminent (such as when a cyclone appears certain to strike); or when such an event (such as an earthquake) has already occurred. NDC, in consultation with other committees, as appropriate, will decide when this stage is to be implemented and will announce it accordingly. If circumstances so demand, provincial or other committees may make their own decisions to implement this stage. All relevant organizations and personnel are then to act as required by the plan, under the direction of NDC.

- Within standard operating procedures there would be a callout procedure to inform individual members of resource organizations that they were required for duty in their disaster roles. The transmission/receipt system required to implement this procedure would need to be as "fail-safe" as possible and the means used would depend on local circumstances and capability (e.g., some organizations might have mobile radios; others might have to rely on a siren system; others might be dependent on the telephone; others, again, might have to use word-of-mouth).

- Equipment and vehicles would be readily available for manning, deployment, and use.
- There would be enough warning time to enable resource organizations:

  - to take any necessary precautionary measures laid down in counter-disaster plans and standard operating procedures, and
  - to be operationally ready in time to meet the threat.
- There would be timely activation of the EOC system so that no delays from this source would be imposed on resource organizations.

Obviously, with certain resource organizations and/or in specific disaster circumstances, additional factors might apply. However, those stated above are likely to be the most general ones.

### 21. *Factors Affecting Best-Case Situations*

Factors that might affect the foregoing best-case situation and which might downgrade efficiency in the activation of resource organizations are:

- Warning factors (e.g., in no-warning situations it could be expected that activation, and therefore deployment for response operations, would be delayed; similar effects would arise from delays or breaks in the warning system).
- Loss or damage factors (e.g., loss of or damage to communication systems, equipment, etc. caused by disaster impact in no-warning situations).
- Physical access and movement factors (e.g., in short warning situations, such as flash flooding, which might prevent mobilization of personnel and vehicles).
- Preparedness factors (e.g., inadequacies in plans, standard operating procedures, or other preparedness arrangements).

## Graduated Response

22. Obviously, there are benefits to be derived from the implementation of plans and standard operating procedures by stages. This concept can be taken a stage further in the sense of applying a form of graduate response to operations; in other words, managing resources so that some capability is kept in reserve. The application of this system obviously depends very much on particular circumstances.

23. However, there is often a general tendency, immediately after disaster impact, to throw maximum resources into operation. Sometimes this is justified, for instance:

- to save lives by using a maximum search and rescue effort, or
- to prevent floodwaters from entering a vital area, or
- to contain a major fire problem.

In other circumstances, a graduated use of resources is more appropriate and this especially applies when response operations are likely to be prolonged. In this regard, it is worth bearing in mind that community self-help can often be used to bolster resource organizations and thus spread the efforts of the latter over longer periods. For instance, if a major welfare center program is obviously going to be prolonged, specialist or professional resources can be eked out by using community members in appropriate roles.

In any case, whether self-help elements are used or not, it is beneficial to use resources on a graduated response basis because this usually means that fresh personnel can be brought into operations at appropriate intervals. This helps maintain efficiency levels in dealing with the various tasks and, therefore, lifts the effectiveness of overall response.

24. It is suggested that this aspect especially warrant the attention of disaster managers who are responsible for direction/coordination and those who manage resource organizations.

## Commodity Management

25. In considering use of counter-disaster resources, it is natural to focus primarily on resource organizations because these tend to constitute the main sources of equipment, vehicles, and trained personnel. However, an important part of resource management is concerned with commodities, such as fuel, food, emergency materials, and so on. Four major considerations in this regard are:

- *Planned management of government stocks*
  Some countries are able to maintain stored reserves of special disaster-relevant items, such as tents and generators. However, many countries find themselves having to rely on existing in-country working reserves, for instance:
  - stockpiles of medical supplies, which are rotated through normal usage and resupply;
  - reserves of building materials held for government purposes. These sources of supply need to be reviewed for their suitability and emergency availability, and arrangements made for their possible use under disaster circumstances.

- *Earmarking of private sector supplies*
  It is also necessary to know, in time of disaster, the types, amounts, locations, and broad availability of disaster-relevant commodities held within the private sector, so that these may be quickly purchased by government when required. Often, international disaster assistance funds are available for this purpose. A broad assessment of these private sector resources therefore needs to be made and maintained by the appropriate disaster management authority.

- *Supplies relevant to resource organizations*
  It is useful if resource organizations can maintain a record in their standard operating procedures of parallel resources within the private sector (see Appendix to this chapter). These parallel resources usually tend to consist of items such as vehicles and equipment but including appropriate stores is worth considering. This action should not, however, duplicate action taken under the foregoing category (i.e., the earmarking of private sector supplies).

- *International assistance commodities*
  These also comprise a major aspect of commodity management, involving reception, off-loading, cataloguing, storage, security, and distribution (see also Chapter 9 on international disaster assistance). Disaster management authorities would have always to be well advised to try to assess, as accurately as possible, the requirements involved.

# Annex to Chapter 13

## SUGGESTED FORMAT FOR STANDARD OPERATING PROCEDURES

GOVERNMENT OF EXLAND
NATIONAL DISASTER PLAN
MINISTRY OF XXX
STANDARD OPERATING PROCEDURES

1. **Reference**
   These procedures are to be used in conjunction with the National Disaster Plan.

2. **Purpose**
   The purpose of these procedures is to provide a quick and easy reference to the actions to be undertaken by the Ministry in times of disaster.

3. **Preparedness Action**
   Ensure general readiness of Ministry to respond to disaster.
   Ensure National Disaster Plan is accurate and appropriate from Ministry's own viewpoint.
   Ensure that Standard Operating Procedures are current and that staff are aware of their responsibilities.
   State any other aspects appropriate to Ministry's preparedness.
   State if senior staff are represented on National Disaster Council or any other senior disaster management body.

4. **Operational Tasks**
   State operational tasks as laid down in National Disaster Plan, with any amplifications necessary to ensure that tasks are clear to Ministry staff.

5. **Resource Available**
   List resource suitable for and likely to be available for disaster purposes. It may be useful to list these under the headings of:
   Personnel – with skills and numbers available
   Equipment
   Facilities

6. **Supplementary Resources**
   List these under categories such as:
   Trained persons with the community who might be called upon to reinforce Ministry staff.
   Equipment which could be used to supplement that of the Ministry.
   International resources known to be, or likely to be available.
   Contact arrangements, as relevant, need to be included in this section.

7. **Activation Guidelines**
   List the most likely sources of information on which the Ministry would be activated (e.g., Police and/or National Disaster Council).
   List activation system within Ministry.

8. **Senior Department Supervisors**
   List names, duties, and telephone numbers of supervisors who are responsible for activation and initiation of action by Ministry staff.

9. **Direction and Coordination**
   Outline system of direction and coordination which would apply to Ministry's actions, along lines indicated below:

   *In No-Warning or Short-Warning Situations*
   Permanent Secretary (or designee) would authorize emergency response and exercise direction of departmental activities.
   If situation likely to become prolonged (as judged by the National Disaster Council):
   - Task allocation (with priorities) and coordination would be done by NDC/Operations Control Group.
   - Direction of departmental activities would be by Permanent Secretary.

   *In Long-Warning Situations*
   Task allocation (with priorities) and coordination would be done by NDC/Operations Control Group.
   Direction of departmental activities would be by Permanent Secretary.

10. **Operational Checklists**

    *Stage 1 - Readiness (or Alert)*
    Specify action such as:

Warn staff
Action, as necessary, to secure Ministry property equipment and facilities.
Check supplies and equipment for use during disaster.
Warn supplementary trained persons in community (if applicable).
Other points relevant to particular Ministry.

*Stage 2 - Stand-by*
Specify action such as:
Place staff on stand-by.
Deploy resources or man vehicles as necessary.
Prepare to carry out operational tasks.

*Stage 3 - Action*
Specify action such as:
Carry out operational tasks as notified by NDC/Operations Control Group.
Maintain communication with Operations Control Group/EOC.
Report tasks' progress and tasks completed to EOC.

*Operation Complete*
Assume normal status, on advice from NDC.
Revert to Preparedness Action, as in Section 3 above.

11. **Administrative**
Maintain disaster log and necessary records to facilitate post-disaster administrative and accounting action.

12. **Stand Down**
As directed by Permanent Secretary, on information from NDC.

13. **Debriefing**
Carried out on instructions of Permanent Secretary, as soon as possible after completion of operations. Submit report to National Disaster Management Office, for NDC.

Date Permanent Secretary

Note: These Standard Operating Procedures can be adapted for use by other agencies and at various levels of government.

# PART II

# Long-Term Measures

*Chapter 14*

# Prevention

## Purpose

1. The purpose of this chapter is to outline the following major considerations which apply to disaster prevention:

- The need to consider disaster prevention;
- Problem areas in prevention;
- Positive approaches toward prevention; and
- Resources relevant to prevention.

## The Need to Consider Disaster Prevention

2. For ease of reference, the definition of prevention used in this handbook is restated here. It covers those measures that are aimed at impeding the occurrence of a disaster event and/or preventing such an occurrence having harmful effects on communities. Constructing a dam or levee to control floods is an example of a preventive measure. Controlled burning off in a bushfire-prone area, prior to the high fire-risk season, is another example. Under this definition, it would appear prudent for disaster managers to consider prevention in the context of the following aspects:

- National development planning;
- National disaster management policy;
- Disaster legislation;
- Counter-disaster planning; and
- Special disaster-related circumstances in which particular national assets may be at risk.

It is with these possibilities in mind that the remainder of this chapter is devoted mainly to examination of some of the problem areas in prevention and some of the positive approaches which might be taken.

## Problem Areas in Prevention

### 3. *Traditional Outlooks*

There may be long-standing acceptance of hazards by governments and communities. For instance, a nation may have lived for centuries with a recurring major flood problem. Therefore, the need for preventive measures is not recognized.

### 4. *Cost*

The cost of some preventive measures can be very high; for instance, large public works of similar engineering projects; thus, they tend to be ruled out, perhaps without a detailed analysis of cost–benefit and other factors.

### 5. *Other National Priorities*

Higher priorities given to other major national programs sometimes totally preclude consideration of preventive measures. For example, nations usually prioritize programs such as medical and health, education, economic development, and so on. In addition, considerations affecting disaster may be omitted altogether from such plans. Thus, measures of prevention do not receive adequate or appropriate attention in national planning. In fact, some national development projects may actually increase disaster risks, rather than help prevent them.

### 6. *Political Aspects*

Political motives may sometimes have an adverse effect on disaster prevention. For instance, measures of prevention may upset the interests of sections of the public by encroaching on land, property, or other aspects. The possible political risk to a particular government or political party may not, therefore, be acceptable. Also, the spread and increase of population may expose more people to disaster risks because they may be obliged to live in, say, flood-prone areas on low-lying islands. However, to compel such people to move, as a measure of disaster prevention, may not be feasible for political and/or practical reasons.

### 7. *Development Problems*

Normal development may cause problems related to prevention. For example, the safety considerations which apply to an airport may be affected by a number of issues. Possibilities are:

- The surrounding population may increase significantly, putting more people at risk in the event of a major air accident.
- The operating characteristics of new generations of aircraft may increase the real or perceived risks to the surrounding communities.

Pressure for preventive measures in these cases (e.g., for the closure of a particular airport) may well come from action groups or whole communities. However, such preventive measures may not be economically feasible and are therefore ruled out or indefinitely postponed.

Other aspects of development may produce the need for measures of disaster prevention. For instance, a large-scale release of hazardous chemicals may constitute grave risks to large numbers of people, livestock, and the environment. Similarly, a fire in a major multistorey building may threaten the lives of hundreds of people. In both these cases, strict safety measures are needed and they form the basis of disaster prevention. It must be remembered, however, that high standards of training for staff are usually vitally important and that public awareness and common sense are necessary to support preventive plans and procedures. Thus, it may be difficult, in any case, to guarantee required standards of disaster prevention.

### 8. *Balance in Disaster Management*

Lack of appropriate preventive measures may often throw heavier loads on other aspects of disaster management when disaster occurs. This may especially apply to response operations and recovery action. Also in post-disaster review or analysis, high priorities tend to be given to restoration of infrastructure and to aspects such as rehousing programs. Consideration of future disaster prevention usually comes as a lower priority if, indeed, any consideration is given to it at all. There is also the relevant point that over many years, priorities in international disaster assistance tended to be given to relief and recovery measures, rather than preparedness, mitigation, and prevention.

### 9. *Public Apathy*

Public apathy toward and/or awareness of disaster risks may mean that there is no pressure on governments to consider disaster management measures, least of all prevention.

## Positive Approaches Toward Prevention

10. The problem areas in disaster prevention tend to require various forms of countermeasures. However, the nature of prevention is such that the measures

involved usually need to be implemented from senior levels of government. The population of a single community or area is unlikely to be able to institute, for example, a major flood prevention project (though such populations can produce pressure through action groups and other means). The possible approaches suggested below therefore have these constraints in mind.

## 11. *National Policy*

There is need for a clear and comprehensive national disaster policy which addresses the total disaster management spectrum, including consideration of all aspects of prevention. Within this policy there must be a readiness on the part of government to institute preventive measures (especially where these affect the national interest) regardless of their unpopularity.

## 12. *Legislation*

If necessary, there should be resort to legislation to implement measures of prevention; for example, mandatory building codes.

## 13. *Assessment and Monitoring*

There should be adequate assessment and monitoring of disaster hazards and vulnerabilities, so that the need for prevention is accurately identified and defined. This should lead to accurate evaluation of all reasonable disaster prevention projects. In this regard, it is especially important to achieve sensible cost–benefit comparisons; for example, whether, by instituting preventive measures, the nation and community is going to gain more (bearing in mind project costs), as against the losses which may arise if nothing is done.

## 14. *Planning and Organization*

Full and appropriate consideration of all disaster aspects should be given within the national development plan, including the immediate and long-term cost–benefit implications of taking or not taking preventive action.

In this context, the establishment and maintenance of a permanent disaster management section or center can play a vitally important part because on behalf of government, the section/center should keep a constant watch on disaster management. Thus, it is able to identify the need for preventive measures, whenever such need may arise. It is then the responsibility of the section/center to advise government on needs in the disaster prevention field and the priorities which should apply.

Furthermore, there should be insistence by the disaster management section/center, on behalf of the government, that an effective post-disaster review is undertaken after all major disaster events. This review must include advice to government on whether, as a result of the particular disaster, further preventive measures are warranted.

### 15. *Public Awareness and Education*

Public awareness and education programs should ensure, among other things, that disaster-prone communities are kept aware of the risks and vulnerabilities that may apply to them. In this way, communities are likely to support the need for sensible disaster prevention, if this becomes necessary.

### 16. *International Assistance*

The maintenance of a continuous dialogue with international assistance agencies can also be of use. Such a dialogue helps ensure that any proposals concerning disaster prevention can be evaluated and submitted to appropriate assistance agencies.

## Resources Relevant to Prevention

17. It is important for disaster management authorities to recognize that, usually, the consideration, decision making, and measures of implementation that apply to prevention are far-reaching. Some of the main resources which are likely to be involved are given subsequently:

*Resources to achieve identification and analysis of disaster threats*

- academic institutions;
- research establishments;
- technical authorities;
- scientific programs; and
- government departments and agencies which may have major public safety responsibilities (e.g., those dealing with land, sea, and air transport systems);
- private sector authorities concerned with programs and projects which may generate potential disaster threats (e.g., companies producing hazardous chemicals);
- industrial safety organizations; and
- international assistance agencies in connection with development projects of various kinds.

*Resources to assess the potential need for and possibilities of implementing programs of disaster prevention*

- government departments, organizations, and agencies;
- national planning authorities;
- disaster management authorities; and
- co-opted specialist institutions.

*Resources to implement disaster prevention programs and measures*

- government departments (especially those possessing technical capability and expertise);
- private sector companies and contractors;
- NGOs;
- military forces;
- international disaster assistance agencies; and
- disaster management authorities.

*Resources to support disaster prevention activity*

- NGOs;
- media;
- general public; and
- educational authorities.

*Chapter 15*

# Mitigation

## Purpose

1. The purpose of this chapter is to outline major considerations which apply to disaster mitigation.

2. There are various definitions of disaster mitigation. For the purposes of this handbook, mitigation is defined as involving:

> Measures aimed at reducing the impact of a natural or man-made disaster on a nation or community.

With this definition, the basic assumption is that, while it may be possible to prevent some disaster effects, other effects will obviously persist. The concept of mitigation recognizes this and maintains that the application of certain measures (usually in the form of specific programs) can moderate or reduce disaster effects.

3. Many factors that apply to prevention also apply to mitigation. However, in the interests of clarity and ease of reference, this chapter is intended to be as self-contained as possible. Thus, any repetitions from Chapter 14 are made intentionally.

## Acknowledgments

4. A good deal of the information contained in this chapter has been taken from the Technical Background Paper of *Disaster Mitigation in Asia and the Pacific* (1990), prepared by the Asian Disaster Preparedness Center of Asian Institute of Technology, for the Asian Development Bank. The authors of this paper are Dr. Ian Davis and Professor S.P. Gupta. This contribution to the handbook is gratefully acknowledged.

## Contents of the Chapter

5. The contents of the chapter cover:

- Previous references to mitigation in the handbook;
- Guiding principles of mitigation;
- Examples of mitigation measures;
- Problem areas in mitigation;
- Requirements for effective mitigation;
- Major mitigation components;
- Resources relevant to mitigation; and
- Formulation and implementation of mitigation programs.

## Previous References to Mitigation

6. Other chapters of the handbook contain references to mitigation as follows:

- The introductory notes illustrate the place of mitigation in the disaster management cycle.
- Chapter 1, on the significance of disaster, stresses the need for disaster mitigation from both a global and national viewpoint.
- Chapter 3 refers to mitigation as a main element in national disaster management policy.
- Chapter 5 contains a further reference to the place of mitigation in the disaster management cycle.
- Chapter 7, on disaster legislation, refers to the inclusion of mitigation in overall national planning.
- Chapter 9 quotes mitigation as a possible area for pre-disaster assistance from international resources.
- Chapter 12 reiterates the need to include mitigation programs in counter-disaster planning.
- Chapter 14 refers to the practice, in some cases, of using the term prevention to include also mitigation.
- Chapter 16 refers to prevention and mitigation as usually being geared to major policy decisions at government level, whereas preparedness tends to be more locally and institutionally oriented.
- Chapter 20 emphasizes the need for post-disaster review committees to address problems concerned with mitigation.
- Chapter 22 emphasizes the need for communities to have a broad understanding of government counter-disaster responsibilities, including mitigation.

- Chapter 23 mentions mitigation as an area of modern disaster research.
- Annex A covers vulnerability analysis aspects of mitigation.

## Guiding Principles of Mitigation

7. The following principles are widely recognized as providing a valuable guide to disaster mitigation.

A. INITIATION

- Disasters offer unique opportunities to introduce mitigation measures.
- Mitigation can be introduced within the three diverse contexts of reconstruction, new investment, and the existing environment. Each presents different opportunities to introduce safety measures.

B. MANAGEMENT

- Mitigation measures are complex and interdependent, and they involve widespread responsibility. Therefore, effective leadership and coordination are essential to provide a focal point.
- Mitigation will be most effective if safety measures are spread through a wide diversity of integrated activities.
- "Active" mitigation measures that rely on incentives are more effective than "passive measures based on restrictive laws and controls."
- Mitigation must not be isolated from related elements of disaster planning such as preparedness, relief, and reconstruction.

C. PRIORITIZATION

- Where resources are limited, priority should be given to the protection of key social groups, critical services, and vital economic sectors.

D. MONITORING AND EVALUATION

- Mitigation measures need to be continually monitored and evaluated in order to respond to changing patterns of hazards, vulnerability, and resources.

E. INSTITUTIONALIZATION

- Mitigation measures should be sustainable to resist public apathy during the long periods between major disasters.
- Political commitment is vital to the initiation and maintenance of mitigation.

## Examples of Mitigation Measures

8. For purposes of illustration, some simple examples of mitigation measures are:

- Strengthening buildings to render them more resistant against cyclones, floods, or earthquakes.
- The incorporation of hazard resistance in structures or procedures to be followed in new development projects.
- Planting certain kinds of varieties of crops that are less affected by specific kinds of disaster.
- Changing crop cycles so that crops mature and are harvested before the onset of the flood or cyclone season.
- The adoption of land-use planning and controls to restrict activities in high-risk areas.
- Economic diversification to allow losses in one sector to be offset by increased output in other sectors.

The foregoing activities can be broadly classified into structural and nonstructural mitigation measures, as outlined in paragraphs 14–23.

## Problem Areas in Mitigation

9. From a practical disaster management viewpoint, certain problem areas can apply to mitigation. Some of these are given below.

- There may be long-standing acceptance of disaster risks by governments and communities, who may feel that traditional measures (taken over many years) are adequate. Such measures may include the positioning of population sites and traditional building practices. There may therefore be some built-in reluctance to accept new methods of mitigation.
- Some mitigation measures may be costly; for example, enforcement of building codes is likely to increase the cost of buildings. This in turn may reflect, in various ways, on costs and prices, and may therefore be opposed within the community.

- Higher priorities given to other major national programs (health, education, etc.) may make it financially difficult to implement mitigation programs.
- Political considerations may rule out or restrict mitigation programs. If such programs are extensive and (through land-use restrictions and enforcement of building codes, etc.) unduly interfere with living conditions and standards, government may become unpopular. Thus, governments may not enforce mitigation programs to full effectiveness, in the interests of retaining political power.
- Aspects of modern progress and development may affect mitigation programs. For instance, international standards in various fields may dictate that governments undertake mitigation measures (mainly perhaps in the form of safety measures). In such cases, governments may have little or no choice. This may therefore mean that other desirable mitigation programs will have to take lower priority.
- Lack of insufficient appropriate mitigation measures may have an adverse effect on the ability to cope with disaster situations. For instance, inadequate mitigation measures may cause a significant overload on response operations and result in the latter being only partially effective.
- If counter-disaster planning is inadequate, the effectiveness of mitigation may be seriously reduced. For example, it may be possible to mitigate the effects of a disaster situation by undertaking a precautionary evacuation of people before the disaster strikes. However, if evacuation plans and other counter-disaster arrangements are not already in place, such an evacuation may not be possible. Even worse, if such an evacuation is attempted without adequate plans, the risks to the people concerned may even be increased. Similarly, if planning is inadequate, large numbers of people may, as a mitigation measure, be housed in unsafe communal buildings (schools, churches, etc.) In the past, in some countries, this has led to considerable loss of life because of buildings collapsing during disaster impact.
- In post-disaster analysis and review, insufficient attention may be given to mitigation measures. This can have severe repercussions in future disasters.
- Inadequate standards of community self-reliance and self-help may adversely affect successful mitigation because even elementary precautions (such as ensuring an emergency food supply, or being prepared for evacuation) will not be taken.

## Requirements for Effective Mitigation

10. Areas of need in prevention tend to be the responsibility of senior levels of government and of senior management in the private sector. To some extent, this applies to mitigation, where measures such as building codes and land-use regulations usually emanate from major policy decisions. However, with mitigation, lower levels of government may play a greater part. For instance, precautionary evacuation as a mitigation measure tends to be the responsibility of local government or the local counter-disaster committee. Also, precautionary evacuation may even result from a local community decision or reaction.

11. Requirements for effective mitigation may include some or all of the following:

- A clear and comprehensive national disaster policy which addresses all aspects of disaster management and ensures that mitigation is given proper consideration and priority.
- Adequate assessment and monitoring of disaster hazards and vulnerabilities, so that the need for mitigation measures is accurately identified and defined. Indeed, effective vulnerability analysis is a primary prerequisite for mitigation programs and Annex A deals with this subject in detail.
- Adequate and accurate analysis of all reasonable mitigation projects. In this regard, it is especially important to achieve sensible gain/loss comparisons; for instance, whether by instituting mitigation programs the nation and community are going to gain more (bearing in mind the costs and restrictions involved), as against the losses which might arise if nothing is done.
- Readiness on the part of governments to institute and carry through appropriate mitigation programs.
- Appropriate consideration of mitigation measures in national development plans, including the immediate and long-term cost–benefit implications of taking or not taking mitigation action.
- A basis of organization and planning centered on a permanent disaster management center or section. The existence of such a section is vitally important in the overall disaster management sense because, on behalf of government, the center/section should keep a constant watch on disaster management. Thus, in coordination with other agencies, it is able to identify the need for various mitigation measures as they may arise. It is then the responsibility of the center/section to advise government on needs for mitigation programs, and the priorities which should apply.

- Insistence by the disaster management center/section (on behalf of government) that an effective post-disaster review is undertaken after all major disaster events. This review must include advice to government on whether, as a result of a particular disaster, mitigation measures are adequate or whether additional measures are needed.
- Recognition that mitigation measures may originate from all levels of government, not only from the national level. This is important because, for instance, the "disaster front" is usually at the local government (or community) level. Thus, from this level, the need for mitigation measures may be more obvious than from higher levels.
- Specialist programs which may assist in the development of large-scale mitigation measures; for instance, agricultural programs which assist farmers and others in mitigation of crop losses.
- Adequate public awareness and education programs, in order to assist communities in playing their appropriate part in mitigation measures.
- Support for traditional measures of mitigation, where these may be used in the overall disaster management sense.
- Support, also for the development of self-reliance and self-help at the community level because these aspects can often provide useful support for mitigation concepts.

12. In coping with problem areas, a participatory approach that involves governments, communities; and others concerned may be invaluable. Such an approach can particularly assist in identifying needs, establishing practical solutions and a combined sense of responsibility in program implementation.

## Major Mitigation Components

13. Bearing in mind the mitigation problem areas and requirements outlined above, certain major components or activities generally apply to mitigation programs. These are covered below under the headings of nonstructural mitigation and structural mitigation.

### *Nonstructural Mitigation*

14. *Legal framework* – Generally speaking, existing disaster-related legislation tends not to place enough emphasis on mitigation. In establishing or reviewing such legislation, therefore, it may be advantageous to ensure that mitigation requirements are adequately covered. Land-use planning and the application of building codes provide some legal basis for successful mitigation. However,

both these aspects tend to fall short of full effectiveness unless they are rigidly enforced.

15. *Incentives* – Incentives can often provide better inducements for mitigation than legal impositions. Government grants or subsidies may help persuade commercial and other institutions to include mitigation measures in their building or reconstruction activities. The provisions of government technical assistance can help toward the same end. Insurance can also provide useful incentives: for instance, insurance companies may be persuaded to offer reduced premiums for buildings, once hazard-resistance measures have been incorporated.

16. *Training and education* – If mitigation were to be successful, its requirements must be widely known and understood. Therefore, there is a need to train and educate all those involved, including disaster management officials, construction specialists, and the general public. In this regard, public awareness programs can provide an important foundation by informing people generally of the need for and benefits of mitigation programs (see also paragraph 17 below for more details on public awareness). In a more specific sense, programs of training and education are necessary to ensure that mitigation programs will be supported and property implemented. Four target groups are especially important:

- Public officials who play a vital role in disaster management. Appropriate training modules should be incorporated in their career path training programs and opportunities provided to them to attend specialist courses.
- Technical students whose professional education should include disaster mitigation courses.
- Small builders and craftsmen who may be given on-the-job training in simple mitigation practices.
- Schoolchildren who should be introduced to simple mitigation measures in the context of environmental studies, natural science, or geography classes. (The Fiji Islands and the Philippines have developed appropriate curricula and materials which may help other countries).

17. *Public awareness* – In addition to general awareness, certain particular areas of public involvement are necessary for effective implementation of mitigation programs. These include:

- A good public knowledge and understanding of local hazards and vulnerabilities.

- Public awareness of the kind of mitigation measures which can be applied.
- Public participation in community preparedness programs.

Governments can substantially assist public awareness of safe mitigation practice by ensuring that their own public buildings (such as post offices, schools, hospitals, government offices) and services are built of high safety standards. This will also help ensure that designers, builders, and engineers gain experience in safe construction and, at the same time, contribute to a safer environment.

18. *Institution building* – The strengthening of a country's or community's social structure can enhance disaster mitigation capacity. Such strengthening is, however, difficult to achieve. Three possible ways are to extend normal development as follows:

- First, through institution building. Organizations that serve as coping mechanisms can be identified and strengthened. A deliberate effort can be made to increase their institutional capacities and skills, thus enhancing their ability to deal with a crisis.
- Second, through increasing the number of coping mechanisms within a country or community. By developing formal institutions and linking them to outside resources, means are established for intervention and the provision of assistance.
- Third, through encouraging actions that promote cooperation among different groups within society. Such cooperation can considerably reduce the social impact of disasters.

In their development activities, both government and nongovernment agencies should be careful to avoid actions that will further increase or institutionalize a society's vulnerability. It is especially important to identify institutional dependency relationships, particularly those that may be increased in disaster situation, and work to eliminate them. By increasing self-sufficiency, agencies may improve the ability of families and communities to cope with disaster. This can be a mitigating factor and help speed recovery. Strong institutions can play a vital role in various aspects of mitigation such as promoting public awareness programs, training at community levels, and monitoring hazards and vulnerabilities.

19. *Warning systems* (*see also Annex B* ) – Various modern developments have significantly improved the ability of disaster management authorities to provide effective warning of impending disaster. Better warning systems have, for instance, been instrumental in evacuating vulnerable groups, moving

livestock to safety and mobilizing emergency services and resources. These and associated matters are covered in detail in Annex B; however, in the particular context of mitigation, three are underlined here.

- The steps between the issuing of warning and the taking of action by relevant authorities or vulnerable people are critical.
- Evacuation should only be ordered when hazard impact is virtually certain; a force evacuation order for a hazard that does not materialize can destroy public confidence in the warning system and neutralize several years of preparedness planning.
- To the extent possible, the dissemination of warnings should use duplicate systems to ensure effectiveness; for example, radio message backed up by siren warnings, and warning flags backed up by house-to-house visits by local wardens.

20. *Agricultural mitigation measures* – Various measures can be applied in agriculture to mitigate the effects of disaster. These include:

- The planning of shelter breaks, comprised of trees and shrubs, to reduce wind effects.
- Crop diversification.
- Adjustments to crop planting/harvesting cycles.
- Food storage programs to insure against shortage arising from disaster.

## *Structural Mitigation*

21. Nonstructural mitigation measures may need to be complemented by structural measures. In the case of flood-prone areas, embankments, regulators, drains, or bypass channels can be provided, where appropriate, to protect areas from damage by floods. Techniques to mitigate the effects of earthquakes, cyclones, and floods on structures also exist. Structural mitigation measures may apply to both engineered structures and non-engineered structures.

22. *Engineered structures* – Engineered structures involve architects and engineers during the planning, designing, and construction phases. They may include buildings ranging in scale from dwellings to multistorey office blocks, as well as infrastructure such as electricity pylons to dams, embankments, ports, roads, railways, and bridges. While professionals are trained to plan, design, and supervise the construction of buildings and infrastructure to achieve necessary structural safety standards, they may need additional training to incorporate mitigation practices into their design of structures resistant to seismic shock, storm winds, or floods. The application of sound technical principles is achieved through:

- site planning;
- assessment of forces created by the natural phenomena (earthquake, typhoon, and flood);
- the planning and analysis of structural measures to resist such forces;
- the design and proper detailing of structural components;
- construction with suitable material; and
- good workmanship under adequate supervision.

Most countries have building codes for engineered construction. These codes provide general guidelines for the assessment of forces and further analysis, appropriate design methodologies, and construction techniques. If a country does not have a building code which specifies design and construction requirements for earthquake and wind forces, such a code should be formulated as soon as possible, and technical personnel trained in its use and enforcement ensured. The quality of construction is as important as good analysis and design. Good workmanship must be encouraged by appropriate training and supervision to achieve performance.

23. *Non-engineered structures* – Non-engineered structures are those constructed by their owners themselves or by local carpenters and masons who generally lack formal training. Such structures mainly comprise simple dwellings and public buildings, built with local materials in the traditional manner. In some disasters, high casualties and economic losses can be attributed to the failure of non-engineered structures. The improved designs vary according to the many different traditional ways of building that suit various cultures, climates, available skills, and building materials.

Another important aspect of increasing the safety of non-engineered structures is to try to ensure that they are not built on hazardous sites such as steep slopes subject to landslides, floodplains subject to flash floods or river bank erosion, or coastal areas exposed to storm surges. However, people often do not want to leave their traditional homes and the area in which they have been living for generations, even though the location may be hazard prone. Economic pressures may also induce people to settle in hazardous areas. Wherever practical, incentives should be offered to attract people out of hazardous locations; alternately, consideration may be given to substituting appropriately engineered structures where this might be practical and economic, or mitigation measures introduced in non-engineered constructions to enhance their safety.

## Resources Relevant to Mitigation

24. The main resources most relevant to mitigation measures andprograms are given below.

*Resources to achieve identification and analysis of disaster risks, hazards, vulnerabilities, and similar aspects usually include:*

- academic institutions;
- research establishments;
- technical authorities;
- scientific programs;
- departments and agencies concerned with hazards (e.g., meteorological, seismological, hydrological);
- government departments and agencies which may have major public safety responsibilities (e.g., those dealing with land, sea, and air transport systems);
- government departments and agencies which are responsible for production (e.g., agriculture), which may be affected by disaster, and thus have a responsibility for identifying and implementing mitigation measures;
- private sector authorities concerned with programs which may generate potential hazards (e.g., companies producing hazardous chemicals); and
- industrial safety organizations.

*Resources to assess the possible needs for and possibilities of implementing programs of disaster mitigation*

- government departments, organizations, and agencies;
- national planning authorities;
- disaster management authorities; and
- co-opted specialist institutions.

*Resources to implement disaster mitigation programs and measures*

- government departments;
- construction and similar specialists;
- private sector companies and contractors;
- NGOs;
- international disaster assistance agencies;
- disaster management authorities; and
- communities.

*Resources to support disaster mitigation activities:*

- NGOs;
- media;

- educational and training authorities;
- general public.

## Formulation and Implementation of Mitigation Programs

25. The requirements and circumstances for formulating and implementing mitigation programs are likely to differ in various countries. However, the information given above should be of general assistance. Also, application could be made to:

Asian Disaster Preparedness
Center Asian Institute of Technology
PO Box 2754
Bangkok 10501, Thailand

concerning availability of the Asian Development Bank's publication, *Disaster Mitigation in Asia and the Pacific.*

26. In addition, the following considerations are likely to have some general application.

- If possible, a simple broad strategy should be devised to cover foreseeable mitigation requirements. This strategy should contain component programs, with desirable priorities.
- The strategy should be interlocked, as far as possible, with national development planning, environmental considerations, and other disaster management activities.
- A system for monitoring and reviewing the strategy should be introduced and applied.
- Responsibility for overseeing and coordinating mitigation activities should be clearly defined. Normally, this responsibility would be vested in the minister (or ministerial committee) responsible for disaster-related affairs, with delegation down to the NDC. Responsibility for individual mitigation programs should also be clearly defined.
- An annual progress report covering mitigation activities should be required; this should normally be embodied in an annual disaster management report to the Cabinet.
- Mitigation activity should be regarded as a suitable and productive area for international assistance.
- Since many different agencies are likely to be involved in mitigation programs, the national disaster management office or section should

be authorized (on behalf of government) to fulfill day-to-day liaison requirements to ensure coordination of effort.

- For individual mitigation programs, it is likely that a particular government ministry/department can be given the lead role. For instance, a mitigation program to protect and conserve a vital road system should be led by the ministry responsible for roads.
- Mitigation programs should not be regarded as, or be allowed to become, a separate activity. They should be part of an integrated national program.

# PART III

# Major Factors Prior to Disaster Impact

*Chapter 16*

# Preparedness

## Purpose

1. As with other aspects of disaster management, different definitions can apply to preparedness. The definition used in this handbook is as follows:

> Measures which enable governments, organizations, communities, and individuals to respond rapidly and effectively to disaster situations. Preparedness measures include the formulation to viable disaster plans, the maintenance of resources, and the training of personnel.

2. Given the above definition, preparedness clearly merges with several other major facts of disaster management, some of which have been covered at length in preceding chapters. Cross-reference to these major facts is given in paragraph 4 below.

3. This chapter is therefore designed to interrelate with general preparedness measures, such as planning and training, but to focus more specifically on the development and maintenance of preparedness, including its effectiveness prior to disaster impact. Aspects covered are:

- The nature of preparedness,
- Some problem areas in preparedness,
- A summary of preparedness needs,
- Maintenance preparedness levels,
- Fundings,
- Warning aspects,
- Precautionary measures prior to disaster impact, and
- Resources relevant to preparedness arrangements.

## References to Other Aspects of Preparedness

4. For ease of reference, other aspects that are part of overall preparedness, or which interrelate with it, are summarized below.

- Chapter 3 deals with national disaster management policy.
- Chapter 4 covers major requirements for coping with disaster and includes aspects of organization, planning, resource use, and training, all of which have relevance to preparedness.
- Chapter 5 illustrates the place of preparedness in the disaster management cycle.
- Chapter 11 deals with organization, including components that are fundamental to preparedness capability.
- Chapter 12 deals in some detail with plans and emphasizes the importance of planning in relation to preparedness and response.
- Chapter 13 that considers the utilization of resources and which emphasizes, among other aspects, the importance of systematic preparedness.

## The Nature of Preparedness

5. The references included in paragraph 4 above throw an interesting light on the way in which preparedness is interrelated with so many other aspects of disaster management. In fact, it can almost be said that every aspect of disaster management can affect, or be affected, by preparedness. There is, however, another significant factor which concerns preparedness. This is its different nature when compared with prevention/mitigation. Measures of prevention/ mitigation tend to be geared to major policy decisions at government level; they also are usually directed primarily from senior management levels. Preparedness measures, however, tend to be more strongly oriented toward action by individual organizations. The fact that, generally speaking, significant numbers of organizations are involved in preparedness emphasizes that there is more critical requirement for coordination than may be the case with prevention/mitigation.

## Some Problem Areas in Preparedness

6. Because of its diversity and the large number of organizations which it usually involves, preparedness can produce certain problem areas, as indicated below.

### 7. *Organization and Planning*

- Inadequate policy direction for overall disaster management will tend to have adverse effects on the event and standard of preparedness (as it does on other aspects of disaster management).

- Lack of appropriate counter-disaster plans will also result in inadequate preparedness.
- Outdated plans will also tend to affect preparedness standards. Allowing plans to become outdated is, in fact, a common and serious disaster management shortcoming.
- If disaster management organizational structure is inadequate or inappropriate, preparedness measures will unlikely be fully effective.
- Over concentration on response and recovery measures may lead to low preparedness standards. This also tends to be a fairly common failing in disaster management.

## 8. *Resources*

Unless there is a complete inventory of resource organizations, plus clear allocation of roles and responsibilities to those organizations, gaps or overlaps in preparedness arrangements are likely to exist.

## 9. *Coordination*

- Inadequate coordination in disaster management may result in substandard and/or variable levels of preparedness, because departments and organizations may be working to different preparedness criteria and different priorities.
- Friction and/or lack of cooperation between disaster-related organizations can have very bad effects on preparedness. Such problems may arise from inter-organizational rivalry, poorly defined areas of responsibilities, or clashes of personality between senior officials.

## 10. *Readiness*

Without a national or central disaster management section or center to serve as a focal point, it is very difficult to monitor standards of preparedness. For instance, emergency operations centers may become inadequately prepared to respond quickly to the onset of disaster; emergency equipment, such as stand-by power generators, may be allowed to become unserviceable; and emergency communications equipment may not be adequately serviced and tested.

## 11. *Training and Public Awareness*

- Lack of suitable training for disaster management personnel will obviously result in low standards of preparedness.

- Inadequate public awareness and information concerning disasters usually contributes significantly to poor preparedness.

## Summary of Preparedness Needs

12. The maintenance of effective disaster preparedness is a dynamic requirement. Left to itself, preparedness will quickly fade away even to the point of becoming virtually nonexistent. Also, it is always difficult to maintain adequate preparedness levels under circumstances where the disaster threat is low and/or very infrequent.

13. This section of the chapter therefore presents a summary of many different aspects affecting preparedness. The summary is formulated in such a way that it provides a form of ready-reference checklist for disaster management officials. Some advice on the maintenance of preparedness levels is also included.

### 14. *National Disaster Policy*

There is need for a clear comprehensive national disaster policy which covers all aspects of disaster management and which ensures that preparedness is given proper consideration and priority.

### 15. *Disaster Legislation*

Special disaster legislation may be necessary to ensure that preparedness aspects of national policy are adequately covered and implemented. In some countries this has been found necessary, especially to ensure that preparedness measures exist in the private sector and/or within communities.

### 16. *Organizational Structure*

A clear and workable organizational structure must exist so that levels of disaster preparedness are identified. Disaster legislation helps formalize this aspect. There is also need within the organizational structure for some form of national disaster management section or office.

### 17. *National Disaster Management Section (or Office)*

In relation to preparedness, some form of national disaster management section or office is vitally important. As stated in paragraph 12 above, disaster preparedness is a dynamic requirement. Thus, a continuous process of monitoring needs to be applied to the wide range of necessary preparedness activities. This can be done most effectively by a specialist section.

## 18. *Assessment of Preparedness Action*

Adequate arrangements for identifying, assessing, and monitoring the disaster threat are also necessary. In turn, this enables a reasonable forecast to be made of the likely effects arising from disasters. Preparedness is vitally concerned with these effects because they constitute the actual circumstances, events, and problems against which preparations need to be made. Usually, these effects are made and varied. For example, they include:

- casualties,
- damage to and destruction of property,
- damage to subsistence and cash crops,
- disruption of services,
- damage to national infrastructure,
- economic loss, and
- loss of livelihood.

Therefore, preparedness measures to deal with these need to be determined and put in place before disaster strikes.

## 19. *Planning Framework*

If preparedness measures are to be fully effective, they need to be clearly set out in appropriate plans. Such plans usually need to apply at the national, provincial/regional, and local government levels. If preparedness measures are set within this planning framework, responsibilities for them can be clearly and officially defined. This also helps ensure that the measures can be systematically monitored and kept up-to-date.

The production of effective counter-disaster plans usually involves considerable negotiation with resource organizations (e.g., government departments, NGOs), especially to ensure that their capability is used to the maximum extent. This is important because, especially in severe disaster circumstances, the total capability of these organizations is needed to deal with the many operational tasks which arise.

## 20. *Use of Resources*

If available resources are to be used to optimum effect, there must be:

- An accurate and up-to-date inventory of all available resource organizations (e.g., government departments, NGOs, potential international assistance agencies).

- Clear allocation of the roles and responsibilities, which resource organizations are required to undertake during response operations and, where necessary, during the recovery phase.
- Suitable preparedness arrangements within resource organizations to ensure that they are able to fulfill their roles when required. It is advisable that their preparedness arrangements follow common guidelines and that these guidelines be issued by the national disaster management section as part of the general planning process.
- It is also desirable that the capability of resource organizations should be monitored, as appropriate, to ensure that operational roles can be fulfilled. This can usually be done by the organizations themselves but, if necessary, assistance can be provided for the national disaster management section.
- Consideration should also be given to preparedness measures necessary for the reception and use of international resources. These might include arrangements for reception, storage, and distribution of relief commodities; use of assistance teams (e.g., rescue workers, medical personnel, technical teams); fly-in and turn-round of assistance aircraft, including refueling and similar arrangements; and port-handling facilities for visiting sea transport.

## 21. *Coordination of effort*

Disaster preparedness and the response operations to which preparedness essentially applies involves a wide range of activities and organizations. If these activities are to be successfully carried out by the organizations concerned, a system for achieving coordinated effort is clearly needed.

This system is usually provided through the organizational framework. For instance, a provincial disaster committee would normally coordinate activities within its area of responsibility. However, additional coordinating responsibilities may be designated to organizations or individuals, if deemed necessary.

Arrangements to achieve successful coordination of effort must obviously be made, as a part of preparedness, before disaster impact.

## 22. *Operational Facilities and Systems*

Adequate preparedness of the various facilities and systems which are required for response operations is also most important. Such facilities and systems usually include:

- emergency or stand-by communications;
- a warning system, including provision of warning and information to the general public;
- a system for activating the organizational structure and its resource organizations (usually by designating stages such as Alert, Stand-by, and Action);
- emergency operation centers (which form the focal points of information management);
- a system for damage survey and assessment of needs;
- emergency relief arrangements (for food, shelter materials, medical assistance, etc.).

## 23. *Equipment and Supplies*

If stockpiles of emergency equipment and supplies are held, these need appropriate surveillance to ensure their serviceability and ready availability. Emergency equipment needs to be held at the levels where it will be primarily used (e.g., equipment for local self-help teams, such as picks and shovels, needs to be held at the community level). Sometimes safe storage (and thus ready availability) poses some problems. However, these can usually be overcome locally. In one case, the village schoolmaster was made responsible for storing and maintaining emergency stocks and schoolchildren had a part in checking and accounting for them.

Where there is a possibility that equipment and supplies from the private sector may need to be co-opted or requisitioned, preparedness arrangements for this eventuality need to be maintained. (See Appendix to Chapter 13, which gives a proposed format for departmental standard operating procedures. These procedures include provision for earmarking possible supplementary resources likely to be available within the community.)

## 24. *Training*

Training is obviously a most important component in preparedness. If possible, a permanent training system and program is desirable. This should cover not only needs of government officials but also those of NGOs. In addition, training is required for persons (such as volunteers from within the community) who may fulfill or assist in disaster management system or parts of it, and provide valuable preparedness checks. In training, as with so many other aspects, the existence of a permanent disaster management section of office is invaluable.

## 25. *Public Awareness and Education*

An aware, alert, and informed public is a most valuable asset for preparedness. Public awareness programs can be presented in various forms to suit particular circumstances. Events such as a National Disaster Preparedness Day are helpful in promoting and sustaining public awarenessInclusion of disaster awareness in school programs usually has long-term value.

## 26. *Effects of Crisis Pressure*

Disaster impact usually imposes various forms of crisis pressure on organizations and individuals who have to deal with the problems caused. It is difficult to stimulate this crisis pressure in exercises or tests. Therefore, in formulating preparedness measures, it is advisable to take account of crisis pressure and, where possible, to try to compensate for it. This might be done by earmarking additional stand-by personnel, arranging for mobility in switching resources from one area to another, or similar measures.

## Maintenance of Preparedness Levels

27. The maintenance of effective levels of preparedness is a major problem, especially because preparedness will tend to fade away if left to itself. The following examples, which are taken from case study information, are unfortunately all too common.

- Following a severe disaster, efforts were made to improve disaster preparedness, especially by emphasizing measures already included in the national disaster plan. A review of these efforts noted that, despite the obvious need for improvement, government departments did not respond to official reminders concerning preparedness, nor did they submit their state of annual readiness on the specified date, as required by the national disaster plan.

- An international assistance team, engaged on a survey designed specifically to help a disaster-prone country, noted major factors responsible for inadequate preparedness: government apathy, inactivity by the designated national disaster committee, and failure to maintain the viability of disaster plan.

28. In Chapter 12 on plans, it is suggested that certain activities and arrangements help maintain the viability of disaster plans. These are

summarized below because, clearly, measures keep plans active, viable, and help maintain preparedness levels.

- Training activities,
- Exercises and tests,
- Functional and readiness checks,
- Post-disaster review,
- Use of regulations,
- International assistance liaison,
- Public awareness activity,
- Publicity, and
- Education in schools.

29. However, there are usually two particularly key areas which, in most countries, can significantly influence preparedness levels. These are:

- The lead given by the main national disaster management authority (e.g., the NDC or its equivalent)
- Astute use of the outreach capability possessed by most of those NGOs which have a disaster-related capability and role.

30. To use to optimum effect the position and role of the national disaster management authority, the following lines of action need to be pursued. (For purposes of illustration it is assumed that a national disaster plan, with supporting plans, is in place; that an NDC exists; and that the NDC implements its day-to-day activities through a small National Disaster Management Office (NDMO).

*Action program of NDC*

NDC needs to work to a clearly specified action program which includes items such as:

- A regular meeting schedule;
- An annual Disaster Management Implementation Program, which includes maintenance of planning viability, preparedness, training, and so on;
- If the disaster threat is seasonal (e.g., cyclone or flood), some form of special annual emphasis should be given to this (for instance, check of the warning system);
- Periodic check of functional readiness (e.g., stand-by power supplies for broadcasting stations);
- Regular view of NDMO's work program and allocation of priorities;

- Assistance, usually through the NDMO, to resource organizations;
- Issuance of directives, where necessary, to support NDMO's work with other departments and organizations (e.g., the development of departmental standard operating procedures);
- Periodic circulation of disaster management information sheet or newsletter;
- Inclusion of disaster management in the curricula of government training institutes to assist officials in fulfilling disaster-related roles; and
- Other items to suit particular circumstances and time scales.

*Work program of NDMO*

As indicated above, the day-to-day work program of NDMO needs to be supervised by the NDC (probably through its chair) and priorities adjusted as necessary. The main theme of NDMO's program should be to maintain the viability of disaster management during nondisaster times, so that capability to cope with disaster, when it strikes, is ensured. The work program of NDMO therefore needs to monitor:

- Policy and organization,
- Training,
- Public awareness,
- International liaison, and
- Disaster reference and information facilities.

With special regard to preparedness measures, NDMO needs to monitor and promote action on:

- Operational systems and facilities such as: communications, emergency operations centers (including earmarking and readiness of staff), warning and public information, survey and assessment, and disaster assistance supplies.
- Coordination and cooperation between all disaster management agencies, including NGOs and the private sector.
- Maintenance of close liaison with other levels of government; for instance, provincial and local disaster management committees.
- Initiation of activities such as workshops, which help promote awareness and readiness.
- Viability of activation arrangements.

31. In many countries, particularly over recent years, NGOs have played an increasingly prominent role in disaster management, especially in preparedness and response. Many governments have realized that, because of restrictions in their own resources, it is beneficial and effective to make optimum use of the religious and welfare organizations which comprise the majority of disaster-related NGOs, because these have a very strong outreach and contact capability as far as the general public is concerned. In addition, NGOs of this type have strong links internationally. It follows that these two capabilities of outreach strength and international linking provide an excellent basis for promoting community awareness and preparedness. Some lines of activity which have proved successful in this regard are:

- Mobilization of NGOs under some form of umbrella organization for disaster preparedness and response purposes;
- Allocation of preparedness and response tasks to individual NGOs or group of NGOs;
- Promotion of public awareness (e.g., publicizing of disaster plans and self-help arrangements);
- Holding of workshops for NGOs, sponsored and funded by international assistance;
- Production of NGO disaster plans to interlock with national disaster plans;
- Stockpiling of disaster stores and other commodities (e.g., emergency food stocks, tools, cooking utensils, medical supplies); and
- Circulation of international newsheets.

32. In sum, if astute linking and programming can be achieved between the official disaster management leadership and the ongoing activities of NGOs, there is every prospect of developing and maintaining good standards of preparedness. What is more, because of the very large voluntary input from most NGOs, the cost of such preparedness can usually be kept to a minimum.

## Funding

33. Whereas the financial requirement of prevention and mitigation is largely self-evident, this aspect is sometimes overlooked with preparedness. Effective preparedness programs do require adequate budgetary support and, for this purpose, international assistance can often be obtained (e.g., for training, communication, warning systems).

## Warning Aspects

34. Warning has sometimes been described as the critical "hinge factor" in disaster management. In other words, it provides the vital link between preparedness measures and response action. From a preparedness viewpoint, therefore, the following aspects are of key importance:

- The warning system, and its associated procedures, must be clearly defined and written down in plans, standard operating procedures, and all other relevant documents.
- The warning system must be known to and understood by all key government ministers, disaster management organizations and officials, other relevant persons, and the general public.
- The system must possess the capability for receiving international warning, initiating in-country warning, and issuing warning at the community, national, and other government levels.
- The system must also include back-up measures, in case its primary components fail or are damaged.
- Intended recipients of warning (e.g., key persons and organizations and the general public) must possess the means of receiving the warning and must know what action should be taken.
- All facilities and arrangements relevant to warning and the warning system (e.g., evacuation arrangements, shelters, and safe havens) must be in an appropriate state of readiness to react to warning.
- Arrangements for activation and mobilization of resources must be in place to make maximum possible use of any warning period.
- All plans and arrangements for disaster response must make provision for no-warning situations (e.g., for an earthquake or volcanic eruption where no prior warning indicators have occurred).
- The warning system and/or component parts of it must be tested and practiced periodically if it is to function satisfactorily when needed.

35. Further information on warning and warning systems is contained in Annex B.

## Precautionary Measures Prior to Disaster Impact

36. If the warning aspects outlined in paragraph 34 are applied effectively, they should provide optimum opportunity to take precautionary measures before disaster strikes. These precautionary measures are likely to vary in different circumstances and situations; they will also depend considerably on

the type and level of threat. However, because of their importance in terms of community safety and survival, disaster management officials should define and apply those measures that fit local conditions. The examples below illustrate the kind of precautionary measures which can generally be applied.

## 37. *Format of Warning Information*

If precautionary measures are to be effectively applied, they depend significantly on the ability of people, especially the general public, to understand what the warning information means. Thus, warning needs to be disseminated in simple, understandable terms, on which people can take action. Preferably, therefore, plain language should be used rather than some form of code. There are sometimes problems in this regard, for instance:

- Internationally accepted regulations and procedures may require the use of specified terminology.
- Some terminology, though simple and clear in itself, may not adequately convey its full significance of recipients. For example, a person who is informed that winds of 85 kilometers per hour are expected may have no idea what this means in terms of risk, damage, or destruction unless he or she has had previous experience of such conditions.

In cases where some form of code necessarily has to be used (e.g., the ringing of bells or the sounding of sirens), there should be no doubt about the meaning. Where possible, therefore, disaster managers need to take local precautions to ensure that warning information can be correctly interpreted. As an example, some countries have devised simple scales of interpretation for wind velocities. These scales can help assess such velocities by what they do to smoke patterns or the bending of trees.

## 38. *Official Government Action*

Some governments feel that by taking official precautionary action themselves (e.g., by closing government offices and institutions) they can help persuade the general public to do likewise. There can, however, be problems in this regard unless care is exercised. A case is known where, during a warning period, government employees were advised by official broadcast to go home. Employees in the private sector therefore did likewise. Subsequently, private sector employers asked the government for compensation to cover lost work output because the warning had applied specifically to government employees and not to those in the private sector.

## 39. *Power Supplies*

In many disaster circumstances, it becomes necessary to cut off main power supplies in the interest of safety. During a precautionary period, therefore, emergency generators should be checked and ample fuel supplies for them ensured. The effect of restricted power supplies on disaster management operations usually needs to be countered in longer-term preparedness measures (e.g., in standard operating procedures). If power supplies are not cut off during disaster, there may be a high risk of loss of life through contact with broken live wires. If this possibility arises, it is usually necessary to inform people accordingly.

## 40. *Transport*

A warning period may also be used for ensuring that essential vehicles are fuelled and, if necessary, dispersed. Also, in island and coastal communities, boats should be fuelled and held, as far as possible, at safe moorings.

## 41. *Crops*

Under some circumstances, it may be advisable to undertake the emergency cutting and collection of crops, rather than risk total damage or loss. This might particularly apply in pre-cyclone conditions or during the slow onset of flooding.

## 42. *Households*

In an aware community, general and seasonal measures are expected to be undertaken as part of overall preparedness. In bushfire-prone areas, for example, various precautions can be taken prior to the onset of high-risk season. These may include general clearance of debris and other fire-prone material, emergency readiness of water pumps, and so on. Similarly, in cyclone-prone regions, general clean-up is a useful precaution, as is checking of roof ties, if these are in use.

Nonetheless, many precautions can also be taken following the receipt of warning of impending disaster impact. These might include:

- Safe storage of emergency food and water supplies;
- Shuttering, boarding-up, or taping windows;
- Broadcasting reminders of what is likely to happen during disaster impact, plus relevant advice (for instance, where to go if evacuation

is necessary). Reminders also of precautions to be taken after impact when there may be danger from fallen power lines, broken glass, weakened structures, and undermined trees.

### 43. *Preparation of Precautionary Information*

Obviously, information concerning precautionary measures needs to be readily available as soon as warning is received. This information therefore needs to be included in disaster plans and, where appropriate, standard operating procedures. The originators of warning (e.g., meteorological and geological authorities) and disseminating agencies (e.g., broadcasting stations) also need to have available the necessary message formats for warning information. It also needs to be made clear in plans and operating procedures that authorities are officially authorized to originate and disseminate warning and other information, so that the risk of false alarm is kept to the minimum. In addition, it is necessary to ensure that information is disseminated in all relevant languages and dialects.

## Resources Relevant to Preparedness Arrangements

44. The main resources relevant to developing and maintaining preparedness arrangements are given below.

*Resources for identifying, analyzing, and advising on the disaster threat:*

- academic institutions;
- research establishments;
- technical and scientific authorities;
- specialist agencies, such as meteorological and geological departments;
- disaster management authorities; and
- records and reports of previous disasters.

*Resource organizations that need to undergo training and other forms of preparedness in readiness for response operations:*

- government departments such as public works, medical and health, transport, communications, agriculture;
- NGOs; and
- emergency services (police, fire authorities, ambulance services, civil defense, armed forces).

*Resources capable of implementing training and other preparedness measures:*

- NDMO
- disaster management authorities

- government training sections;
- NGOs; and
- overseas training institutions.

*Resources to support disaster preparedness activity:*
- NGOs;
- commercial and industrial companies;
- educational authorities;
- media; and
- general public.

# PART IV

# Response to Disaster Impact

*Chapter 17*

# Response

## Purpose

1. In this handbook, the following definition is applied to response.

> Response measures are those which are taken immediately prior to and following disaster. Such measures are directed towards saving life and protecting property and to dealing with the immediate damage caused by the disaster.

2. From the above definition, it is clear that the scope of response is usually extensive and that its success must depend vitally on good preparedness. The effectiveness of response also has a considerable bearing on subsequent recovery requirements and activities.

3. Response operations usually have to be carried out under disruptive and sometimes traumatic conditions. Often, they are difficult to implement and they tend to make heavy demands on personnel, equipment, and other resources. Thus, without a sound basis of planning, organization, and training, response operations are unlikely to achieve optimum success.

4. The purpose of this chapter is to outline the major considerations which apply to response, with particular reference to the following aspects:

- Important characteristics of response,
- Some problem areas in response,
- Requirements for effective response,
- Follow-on from response operations,
- Human factors in response, and
- Resources relevant to various aspects or response.

## Reference to Other Aspects of Response

5. Various references to response have been made in preceding parts of this handbook. The main such references are given below.

- Chapter 2 contains lists of the general countermeasures required to deal with various types of disaster threat.
- Chapter 3 contains various references to the place of response in the national disaster management policy.
- Chapter 4, which covers major requirements for coping with disaster, discusses major aspects of organization, planning, resource use, training and specialist skills, all of which bear on response capability.
- Chapter 5 refers to the place of response in the disaster management cycle.
- Chapter 8 contains information on counter-disaster resources and the roles they may play during response to disaster.
- Chapter 9, on international disaster assistance, contains references to use of international resources in response operations.
- Chapter 10, which deals with leadership disaster, refers to leadership in direction of response operations.
- Chapter 11 discusses various organizational aspects which affect response capability, organizational structures, and response components.
- Chapter 12, which deals with plans, contains various references to the interrelationship between planning and response.
- Chapter 13 covers the utilization resources, primarily in the response mode, and contains suggested standard operating procedures for resource organizations.
- Chapter 16 deals with preparedness, including many measures which are prerequisites for effective response.

## Important Characteristics of Response

6. Effective response to the impact of disaster is critical mainly to:

- limit casualties,
- alleviate hardship and suffering,
- restore essential life support and community systems,
- mitigate further damage and loss, and
- provide the foundation for subsequent recovery.

7. Certain characteristics typically apply to response effort. These include:

- *The type of disaster*
  Depending on its type, the onset of disaster may provide long warning, short warning, or no warning at all. This will obviously influence the effectiveness of activation, mobilization, and application or response effort.

- *The severity and extent of disaster*
  This represents the size and shape of response problem and particularly affects aspects such as:
  - the ability of response effort to cope with the problem;
  - the urgency of response action and the priorities which are applied;
  - exacerbation of disaster effects if appropriate action is not taken; and
  - requirements for external assistance.

- *The ability to take pre-impact action*
  If warning time and other conditions permit pre-impact action to be taken (in the form of evacuation, shelter, and other protective measures), this may have a major effect on the success of response overall.

- *The capability for sustained operations*
  A frequent requirement of response operations is that they must be sustained over a long enough period to be fully effective. Several factors are involved here, including resource capability, management, community self-reliance, and international assistance.
  However, the capability to sustain operations, relative to potential threats, is a disaster management objective which needs to be carefully addressed both during preparedness and response action itself.

- *Identification of likely response requirements*
  An important characteristic of response is that it is generally possible to identify beforehand the kind of response action which is likely to be needed for any particular disaster. As indicated in Chapter 2, on the disaster threat, the effects likely to emanate from individual disasters are well established. Thus, the required response actions are also identifiable. This represents a considerable advantage in disaster management terms in that it is possible to plan and prepare

for well-defined response action in the face of potential threats. This, again, constitutes a tangible objective for disaster management.

8. It is suggested that an assessment of response needs, in the light of the foregoing similar factors, has useful application to most circumstances.

## Some Problem Areas in Response

9. Some of the major problem areas relevant to response are summarized below. These examples have been taken from a number of different factual circumstances and will hopefully, therefore, be of use to disaster management officials.

10. *Background Factors*

These may particularly apply to preparedness, for instance:
- lack of adequate policy direction,
- poor organization, and
- inadequate planning.

11. *Inadequate Preparedness*

This can be caused by:
- plans becoming outdated,
- low standards of readiness on the part of resource organizations,
- poor public awareness, and
- disaster of unexpected magnitude.

12. *Warning Factors*

These may include:
- inadequate warning lead time,
- errors in warning systems (e.g., radio broadcast stations) due to effects of disaster impact, and
- failure of people to respond to warning.

13. *Slow Activation of the Response System*

This may be due to:
- warning factors,

- poor system for activation,
- lack of functional readiness (e.g., in emergency operations centers),
- lack of testing and exercising the response system, and
- coincidence with some national event ( e.g., national holiday).

14. *Effects of Impact and Crisis Pressure*

These may include:

- disruption to or loss of communications,
- destruction or delayed availability of planned resources (e.g., transport, relief supplies),
- damage to key installations such as power supplies, emergency operations centers, communications facilities,
- high damage levels generally, and
- loss of key personnel.

15. *Difficulties in Survey of Damage and Assessment of Needs*

These may arise from:

- adverse weather conditions following disaster impact (e.g., post-cyclonic low cloud and heavy rain),
- lack of suitable aircraft for survey purpose,
- difficulties of ground survey (perhaps caused by problems of access and movement),
- inadequate planning and preparation to cover this requirement, which has to cover a number of detailed aspects, and
- loss of vehicles or vessels.

16. *Inaccurate and/or Incomplete Information from Survey*

This can cause serious response problems through inaccurate figures of people who are:

- homeless,
- without food and shelter, and
- in need of medical assistance.

17. *Convergence*

Convergence onto the disaster area or site by large number of people and vehicles can seriously interfere with response operations.

18. *Poor Information Management*

This may arise from a number of aspects, such as:

- gathering and collation of information,
- evaluation of information,
- decision making, and
- dissemination of decisions and information.

19. *Inadequate Relief Commodities*

This may involve essential items, such as:

- food supplies,
- water supplies, and
- shelter materials (tents, tarpaulins, etc.)

20. *Logistics Problems*

These can be caused by shortage of air, sea, and land transport for the distribution of relief supplies and other activities.

21. *Poor Coordination of Response Operations*

This can result from problems involving

- information,
- duplication of effort,
- unwillingness of some private sector organizations to work within a coordinating system, and
- inadequate training of personnel.

22. *Inadequate Public Awareness*

This can cause various difficulties for disaster management authorities when dealing with the requirements of stricken communities, especially if the latter do not understand local plans and arrangements.

23. *Problems with the Media*

If arrangements for dealing with the media are inadequate or unsatisfactory, this is likely to cause problems for the disaster management authorities.

24. *International Assistance*

Response operations may be adversely affected if expected international assistance is delayed, inadequate, or inappropriate. This can arise if effective prior arrangements have not been made with relevant overseas agencies.

## Requirements for Effective Response

25. Wide international experience has shown that effective response depends fundamentally on two factors:

- information and
- resources.

Without these two vital components, the best plans, management arrangements, expert staff, and so on become virtually useless. Bearing this fundamental premise in mind, the major requirements for effective response are summarized below.

### 26. *General Background of Preparedness*

As indicated in paragraph 3 above, the effectiveness of response operations will depend vitally on the general background of preparedness which applies. This includes various aspects of policy direction, planning, organization, and training.

### 27. *Readiness of Resource Organizations*

The readiness of resource organizations (both government and nongovernment) to respond to disaster situations, often at very short notice, is a very important requirement for response operations. Sometimes, failure on the part of only one designated organization may seriously upset the total response effort. However, disaster management authorities do need to bear in mind that the response lead time for resource organizations can differ markedly. This point is referred to in Chapter 8, which deals with counter-disaster resources. Response management needs to take account of, and harmonize differences in, organizational lead times if a balanced response is to be achieved.

### 28. *Warning*

This paragraph overlaps some of the information contained in Chapter 16. This is intentional to avoid the need for undue cross-reference, especially under impending operational circumstances.

As had been emphasized at several points in this handbook, an effective system of warning is vitally important for successful response operations even though there are bound to be some occasions when little or no warning will be available. The main needs for warning are:

- Initial detection, as early as possible, of the likelihood that a disaster will occur.
- Origination of the warning process as early as practicable, bearing in mind that false or unnecessary warning must be avoided. In this regard, however, precautions can be built into the warning sequence by ensuring that, where doubt exists, only key officials are initially informed.
- Effective means of transmitting warning information.
- Facilities to receive and assess warning information.
- Response decisions, as a result of assessing warning information.
- Dissemination of response decisions and, as appropriate, broadcast of warning information to the public.

As indicated in Chapter 16, preliminary reaction to warning, before a disaster actually strikes, can save lives and property. This preliminary reaction might include:

- closing of schools, offices, and other public places;
- checking emergency power supplies and similar facilities;
- taking precautions in households to ensure supplies of food; and drinking water.

It is reemphasized that preliminary reaction of this kind usually needs to be planned beforehand and, where necessary, the relevant information passed to disaster-related organizations and the public.

(See also Appendix B for further information on warning and warning systems.)

## 29. *Evacuation*

The evacuation of communities, groups or individuals is a frequent requirement during response operations. Evacuation is usually:

- precautionary (in most cases undertaken on warning indicators prior to impact to protect disaster-threatened persons from the full effects of the disaster)

  or

- post-impact (to move persons from a disaster-stricken area into safer, better surroundings and conditions).

The question of evacuation is a complex one which involves a wide range of factors. It is therefore covered in detail in the annex to this chapter.

## 30. *Activation of the Response System*

For rapid and effective response, there usually needs to be a system for activating disaster management officials and resource organizations. It is useful to implement activation in stages. These might be Alert, Stand-by, and Action, as suggested in Chapter 12. The benefit of this arrangement is that if, after the initial warning, the disaster does not materialize, activation can be called off. Thus, full mobilization of resources can be avoided and the minimum of disruption is caused to normal life. It is advisable for government departments and other resource organizations to work to this system of stages in their own internal plans.

## 31. *Coordination of Response Operations*

Coordination of the action taken in response operations is very important. Good coordination ensures that resource organizations are utilized to best effect, therefore avoiding gaps or duplication in operational tasks.

Appropriate emergency operations centers are essential for achieving effective management and accurate decision making.

Also, appropriate disaster management committees (usually at national, intermediate, and local government levels) are necessary to ensure that, as far as possible, there is overall coordination in decision making and in the allocation of tasks.

## 32. *Communications*

As with all aspects of disaster management, good communications are essential for effective response. Also, since communications may be adversely affected by disaster impact, reserve communications (with their own power supplies) are a necessary part of response arrangements. The value of solar-powered communications, especially under severe disaster conditions, can be considerable.

## 33. *Survey and Assessment*

*(See also Appendix C of this handbook)*

It is virtually impossible to carry out effective response operations without an accurate survey of damage and consequent assessment of relief and other needs. To be fully effective, survey and assessment needs to be carefully planned and organized beforehand. It usually calls for:

- survey from the air,
- survey by field teams, and
- accurate reporting from disaster management and other official authorities in or near the disaster area.

In most cases, a general survey needs to be made as early as possible after impact, with follow-up surveys as necessary. Some training is usually required for personnel who are required to carry out survey and assessment duties. This is necessary to ensure the accuracy of information which is collected. The information gathered through survey and assessment is, of course, vitally important for the implementation of immediate relief measures. However, much of the information is also required for the formulation of recovery programs.

## 34. *Information Management*

In the confused circumstances which tend to exist following disaster impact, it is not easy to obtain accurate and complete information. However, without accurate and comprehensive information, it is difficult to ensure that response operations are focussed on the correct tasks, in the right order of priority.

As stated in paragraph 31 above (in relation to coordination) emergency operations centers are essential for effective information management. Especially, EOCs ensure that information is correctly processed according to the proven cycle of:

- acquisition of information,
- information assessment,
- decision making, and
- dissemination of decisions and information.

Therefore, even if there are limitations in obtaining information, the EOC system will make the best use of what is available.

## 35. *Major Emergency Response Aspects*

Following the impact of disaster, there are usually varying degrees of damage to, or destruction of, the systems which support everyday life. Communities therefore need help (usually urgently) to subsist through the emergency phase and beyond. Key aspects of this assistance include:

- *Rescue*
  To rescue persons who may be trapped in buildings and under debris, isolated by floodwaters, or need rescuing for any other reason.

- *Treatment and care of victims*
  To dispose of the dead.
  To render first aid.
  To ensure identification tagging of casualties.
  To identify needs in terms of medical treatment, hospitalization, and medical evacuation; and to deal with these accordingly.

- *Evacuation*
  To determine whether persons need to be evacuated from the stricken area immediately, or whether such a requirement is likely to arise later.

- *Shelter*
  To provide shelter for victims whose housing has been destroyed or rendered unusable. This may involve:
  - making urgent repairs to some housing,
  - issuing tents and/or tarpaulins to provide means of temporary shelter, and
  - accomodating groups of homeless people in community buildings such as schools.

- *Food*
  To organize and distribute food to disaster victims and emergency workers.
  To estimate damage to crops and food stocks.
  To estimate food reserves and available (including unharvested crops).

- *Communications*
  To reestablish essential radio, telephone, telex, and facsimile links.

- *Clearance and access*
  To clear key roads, airfields, and ports to allow access for vehicles, aircraft, and shipping; also to prepare helicopter landing sites.

- *Water and power supplies*
  To reestablish water and power supplies, or to make temporary arrangements for them. Providing potable water is often difficult, particularly in the early post-impact stages. Water-purifying equipment might therefore have to be obtained and/or water-purifying tablets issued.

- *Temporary subsistence supplies*
  To provide supplies such as clothing, disaster kits, cooking utensils, and plastic sheeting, to enable victims to subsist temporarily in their own area, thus helping reduce the need for evacuation.

- *Health and sanitation*
  To take measures to safeguard the health of people in the stricken area and to maintain reasonable sanitation facilities.

- *Public Information*
  To keep the stricken community informed on what they should do, especially in terms of self-help, and on what action is on hand to help them.
  To prevent speculation and rumor concerning the future situation.

- *Security*
  To maintain law and order, especially to prevent looting and unnecessary damage.

- *Construction requirements*
  To estimate high-priority building repair and replacement requirements.

- *Disaster welfare inquiry*
  To make arrangements to handle national and international inquiries concerning the welfare of citizens and residents, including tracing of missing persons.

- *Maintenance of public morale*
  Depending on cultural and other local circumstances, to make arrangements for counseling and spiritual support of the stricken

community. This may involve religious bodies, welfare agencies, and other appropriate organizations.

- *Other requirements*
  Depending on individual circumstances, other requirements, additional to those above, may arise.

## 36. *Allocation of Tasks*

If planning and preparedness have been properly carried out, the majority of response tasks, as outlined in the foregoing paragraph, will have been designated beforehand to appropriate government departments and other resource organizations. For instance:

- Public works department to undertake debris clearance tasks, etc.
- Medical and health department to implement health and sanitation measures.
- Police to maintain law and order, and to assist with control of people and vehicles around the disaster area.
- Red Cross to carry out first aid and other emergency welfare assistance.

The disaster management authority may need to give attention to tasks such as emergency feeding and emergency shelter programs, since these tend not to be in the normal day-to-day schedules of government departments and other organizations.

Priorities for the implementation of response tasks are usually decided by the appropriate level of disaster committee. These priorities may have to be changed frequently and both disaster management authorities and resource organizations need to be capable of accepting and implementing such changes.

## 37. *Availability of Relief Supplies and Commodities*

The ready availability of relief supplies and commodities is an important factor in effective response. After disaster impact, there is usually an urgent need to provide and distribute:

- food,
- drinking water,
- clothing,
- shelter materials, and
- medical supplies and assistance.

Disaster management action therefore needs to cover two main areas:

- Obtaining the various commodities from government stores, emergency stockpiles, commercial supplies, and international assistance sources; and
- Organizing the distribution of these commodities according to the best possible orders of priority.

## 38. *International Assistance Resources*

International assistance resources often play a valuable part in response operations. These resources mainly comprise relief commodities, especially food, shelter, and medical supplies. However, specialist personnel and equipment are also available for damage survey and similar tasks. Disaster management authorities responsible for response operations should also bear in mind that some international agencies and countries hold stockpiles of relief supplies conveniently situated around the world. Access to such stockpiles may be extremely valuable in times of urgent need.

## 39. *Public Cooperation*

Good cooperation between the disaster response authorities and the public is essential if response operations are to be successful. The foundation of such cooperation should, of course, be laid during the public awareness programs which are a necessary part of preparedness. However, disaster response and coordinating authorities should remember that the affected public needs to be kept informed (see also paragraph 35 above). This particularly applies to intended response action and the timing of relief supplies.

If the affected public is not kept as fully informed as possible, rumors and false reports are likely to be started, thus causing problems of cooperation for the response authorities.

## 40. *Media Cooperation*

Disaster, especially major disaster, is news. Consequently, requests for information by local and international media are inevitable. Thus, it is clearly advisable to have well-organized arrangements to deal with this aspect. These arrangements are usually outlined in plans and standard operating procedures, and they are responsibilities of government information and broadcasting agencies (see Chapter 8 on counter-disaster resources). It is important that conditions in the stricken nation be accurately reported internationally and that there should be no misreporting or misrepresentation of international assistance effort. Most disaster events will be superseded by other happenings on the world scene in a fairly short time. Therefore, to avoid possible misunderstandings and misinterpretations, it is important to give media representatives appropriate opportunities to be briefed and to gather information as soon as possible after disaster impact. Delays may lead to some media representatives making their own news, which may not be in the best interests of the affected nation. Good relations with the local media are also important and usually two-way benefits are involved. Not only do the local media benefit from good cooperation from the disaster management authority, but they can also perform valuable services in roles such as warning, evacuation, and public awareness.

It is recognized that during pressurized response operations, disaster management authorities may regard media information as having to take a low priority. However, this should and can be avoided if proper arrangements are in place and appropriate use is made of specialist information staff.

## 41. *Pattern of Response Management*

It is important, especially in the interests of operational coherence, that disaster managers should try to develop and maintain a pattern of management during response operations.

As stated in Chapter 4, which deals with major requirements for coping with disaster, resource management depends on four major factors:

- A capable EOC system;
- A good information picture;
- Effective communication between the disaster management and individual resource organizations; and
- Sensible commitment of resource organizations to operational tasks, bearing in mind their capability and durability.

Given that these factors can be applied, it is useful if the response management authority works to a pattern of:

- Maintaining the best possible information picture (from surveys, situation reports, and other information) concerning the disaster situation and the tasks which may need to be undertaken;
- Establishing priorities for tasks;
- Committing resources to tasks in the most effective manner, bearing in mind that personnel need time for meals and reasonable rest periods.

- Continuously assessing the situation in terms of:
  - tasks completed
  - tasks needing to be undertaken,
  - resources available,
  - possible reinforcement by additional resources, etc.
- Maintaining close liaison with other relevant disaster management authorities (e.g., committees at higher and lower government levels);
- Maintaining close liaison with NGOs;
- Keeping the public as fully informed as practicable; and
- Using self-help from within the community.

### 42. *Period of Response Operations*

Wide international experience indicates that most governments find it expedient to keep the period of emergency response operations down to a fairly limited period. This period usually tends to be 2–3 weeks, after which remaining relief and associated needs are met through the normal systems and processes of government. Undue extension of the emergency is usually regarded as undesirable to avoid:

- overdependence on emergency aid (especially food supplies),
- adverse effects on the local commercial system, and
- unnecessary delay in returning to normal community life.

It may be useful, therefore, for disaster managers to bear this likely time frame in mind in formulating their overall concept of response operations.

## Follow-on from Response Operations

43. From the contents of the foregoing sections of this chapter, it is clear that response operations will usually constitute a short, pressurized period of activity. The major aims of disaster management during this period can be summarized as follows:

- To encounter the initial effects of disaster impact as rapidly and effectively as possible;
- To use all suitable resources in a coordinated manner;
- To provide urgent needs to stricken communities; and
- To rehabilitate, as far as possible, those facilities and systems which are of priority importance to the functioning of the national system and way of life.

44. The emergency response period is therefore a transient one. It does not have a definite cut-off point in terms of national and community requirement. Indeed, from a disaster manager's viewpoint, the period is best regarded as a vital bridge between the shock and disruption caused by disaster impact and the organized process of returning to normal. This means that, following the official ending of the emergency phase, there will be a need to:

- continue certain relief activities (e.g., emergency feeding);
- convert some of these relief activities into more formal types of rehabilitation programs (e.g., it may become necessary for the stricken nation to establish a long-term community assistance program);
- extend some temporary measures (e.g., the emergency clearance and repair of port facilities) into major programs of restoration; and
- assess all post-emergency phase activities and requirement and coordinate them into an overall recovery program.

45. Before the recovery program can be fully implemented (and this may take several months) there tends to be a somewhat blurred period. Many people who have been directly involved in disaster situations have cited this period as the most difficult of all in disaster management. Disaster managers should be aware that this period is likely to arise following most disasters and that it tends to be caused by the following:

- ending of emergency powers which usually apply during response operations;
- transfer of responsibility from the central disaster management authority (i.e., NDC) back to individual government departments;
- necesssary continuance of relief activities by NGOs, whether or not government agencies are involved;
- addition of many disaster-caused problems to the normal workload of most government, nongovernment, and private sector organizations; and
- residual social and psychological problems which are likely to exist within the community following the disaster.

46. It is to overcome this difficult blurred period that some governments have deemed it advisable to use a technical advisory team during the emergency response period. The main purpose of such team (while standing aside from the emergency response operations) has been to identify the strands of relief, rehabilitation, and restoration which emerge post-disaster and bring these strands together for integration into the total recovery program (see Chapter 4). In this way, response operations, emergency relief, initial rehabilitation

programs which constitute the total recovery process. These threads are picked up and continued in Chapter 19 on recovery.

## Human Factors in Response

47. The turbulence and pressures of response operations do not usually allow much opportunity for coping in detail with the trauma inflicted on communities and individuals. Obviously, extreme cases have be dealt with by medical attention and associated counseling, but generally this activity tends to fall within recovery programs. However, disaster management officials should understand that many human factors are involved during response operations. This understanding can materially assist in the assessment of various situations and in decision making. In brief, the main factors tend to be:

- *The plight of disaster victims*
  In extreme disaster circumstances, the plight of disaster victims is severe and traumatic. Shock, personal injury, bereavement, loss, turmoil, and other aspects have a severe effect on the capability of victims to comprehend their circumstances, to realize what is being done for them in disaster management terms and to cooperate in a meaningful and positive way in their own relief and rehabilitation. It is true that extended family members and persons from neighboring community areas will usually be available to provide assistance of various kinds. But severely affected victims themselves constitute a response liability which disaster managers must recognize. In less severe disaster circumstances where community members are capable of providing some coherent self-help, the situation may be more encouraging. However, disaster managers must still recognize that unless levels of community preparedness and experience are high, there may well be problems of organizing self-help in a productive way.

- *The nature of the counter-disaster task*
  The nature of the counter-disaster task also involves human considerations in most circumstances. The nature of response operations can make heavy physical and mental demands on emergency workers. This may well result in lowered functional capacity of resource organizations.

- *The worker/family factor*
  Disaster circumstances will very often result in some emergency workers being separated from their families, with consequent fear

and apprehension on both sides. Indeed, cases are known where, for instance, volunteer firefighters have been engaged in trying to save the community as a whole, while their own families were being put at risk and their homes and properties destroyed. This is clearly a disaster management factor which is difficult to handle and one which can only be mitigated according to local events. However, disaster managers need to be aware of it and to be prepared to resolve it as best as possible.

- *The transfer factor*
  As indicated above and in Chapter 12 on plans, the impact of disaster can affect both victims and emergency workers. However, an additional factor needs to be taken into account. This is the transfer of trauma effects from the disaster victims to the emergency worker. This can particularly apply when emergency workers are inexperienced volunteers and are exposed to disaster trauma for the first time. For instance, a case is known where welfare workers, distributing food to disaster victims, were so badly affected by the latter's suffering that they themselves became unable to continue this important response task.

## Resources Relevant to Various Aspects of Response

48. Disaster circumstances, particularly if they are severe, are likely to require response from the widest possible range of resources. This may even include people who have themselves become victim. Given below are the main categories of resources which are, therefore, applicable to response.

*Resources capable of assisting in the definition of response requirement and the type of operations required to deal with them*

- disaster study and research institutions;
- disaster management authorities; and
- disaster records, especially of post-disaster review and analysis.

*Resource organizations primarily concerned with direction, coordination, and management of response operations*

- national disaster management authority;
- regional or provincial disaster committees;
- special task forces;
- self-contained international assistance teams.

*Resource organizations which provide support for management and direction by providing specialist information and advice*

- meteorological services;
- geological services;
- technical advisory services; and
- specialists in various other fields, as required.

*Resource organizations which carry out allotted tasks in response operations*

- standard emergency services (police, fire, authorities, ambulance services, etc.);
- government departments, organizations, and agencies;
- military services;
- NGOs, including welfare agencies, religious bodies, and a wide range of community services;
- emergency task force teams;
- international assistance teams; and
- community self-help teams.

# Annex to Chapter 17

# NOTES ON EVACUATION

## *Introduction*

1. The evacuation of disaster-affected communities can be one of the most difficult of response operations. As stated in paragraph 29 of this chapter, evacuation is usually:

- *precautionary* (in most cases undertaken on warning indicators prior to impact to protect disaster-threatened persons from the full effects of the disaster); or
- *post-impact* (to move persons from a disaster-stricken area into safer, better surroundings and conditions).

In either case, the decision to evacuate is based on a number of factors, some of which tend to be conflicting. Thus, disaster managers who are faced with the evacuation decision need to have a sound understanding of what is involved.

## *Purpose*

2. The purpose of this annex, therefore, is to outline some major considerations which affect the decision to evacuate individuals, groups, and communities who may be affected by the impact of disaster.

## *The Disaster Threat*

3. One outstanding factor which relates to evacuation is a clear and detailed understanding of the disaster threat. If evacuation decision makers do not fully understand the ramifications of the threat, it is unlikely (if not impossible) that they will make accurate decisions. Also, of course, various aspects of the threat (such as speed of onset and warning lead time) may bear directly and critically on the decision to evacuate.

4. Some relevant aspects of the disaster threat are summarized below:

- *Nature*

  The nature of the threat raises different problems and varying evacuation alternatives. For instance, a simple move to higher ground may be the evacuation solution to moderate flooding. Bushfire, on the other hand, can pose very difficult evacuation problems; in what direction, for example, is it safe to move? In cyclonic conditions, there is nearly always the problem of flying debris, which can pose a hazard to movement and even to life.

- *Severity*
  The severity of the threat may increase or decrease the likelihood of evacuation.

- *Frequency*
  The frequency of the threat affects readiness levels and mental attitudes. A frequent threat may also provide valuable experience in the evacuation decision making.

- *Predictability*
  A high level of predictability in relation to the threat considerably helps in the accuracy of decision making.

- *Speed of onset*
  This usually bears strongly on warning lead time and on time available to implement successful evacuation.

- *Impact effect*
  Very sudden and severe impact (such as that of flash flooding) usually makes organized evacuation difficult. However, it will almost certainly cause self-evacuation.

- *Compound effect*
  Some types of disaster may give rise to compound effects. For instance, heavy rains may cause serious floods and create the need to evacuate. However, rain may also cause landslides or mudslides which impede evacuation movement.

- *Extent*
  An extensive, widespread disaster may create a "nowhere-to-go" situation. This may mean that local safe havens become the only alternative.
- *Past experience*
  Records of past experience in evacuation can offer useful guidelines for the future.

The foregoing brief summary illustrates the importance of the disaster threat in relation to evacuation considerations. In fact, if disaster managers do not understand the threat, it is doubtful if they can begin to think seriously about evacuation.

## General Disaster Management Factors

5. A number of general disaster management factors also bear significantly on the evacuation problem. These are outlined below.

6. *Plans*

The basis of planning is an important factor because it usually provides guidance, at least in general terms, on major aspects of disaster management such as evacuation. For instance:

- *Provincial disaster plans* are likely to define evacuation arrangements in general terms.
- *Local disaster plans* should normally define warning arrangements and specify safe havens and, perhaps, options for evacuation routes.
- *Special plans* may exist in areas which necessitate evacuation.

Therefore, plans usually provide a starting point for tackling evacuation problems.

7. *Preparedness Measures*

In a well-prepared community, readiness measures which likely bear directly on evacuation; for instance, arrangements to prevent (by sand bagging and other means) the breaching of river banks in times of floods. This type of preparedness can reduce the need for evacuation.

8. *Public Awareness*

A community that has a high level of awareness, combined with preparedness, tends to be much more attuned to possible evacuation than a community which is complacent. Also, if good public awareness is prompted, at least in part, by previous experience of evacuation (or even evacuation standby), then future evacuations will usually be easier to handle.

9. *Legislation*

Usually, disaster legislation covers certain aspects of evacuation. These include responsibilities for decision making. The rights of individuals not to be evacuated, if they so choose, are also sometimes covered.

10. *Disaster Management System*

It is extremely difficult, if not impossible, to make an accurate evacuation decision without some form of disaster management system; that is, without the support of:

- Viable communications;
- Provision of warning and other relevant information;
- Facilities for information management (e.g., EOCs);
- Means of disseminating information to the public, and so on.

In other words, the system must provide for two critical needs:

- First, accurate and timely information on which to make the evacuation decision.
- Second, timely, accurate, and clear dissemination to the community of this decision to evacuate.

11. Other background or general factors may apply in differing circumstances, but those mentioned above tend to be the most usual ones.

## *Special Evacuation Factors*

12. In addition to the foregoing general factors, certain special evacuation factors affect the evacuation decision.

13. *Vulnerability*

One such factor is the vulnerability of an area or community. Such vulnerability needs to be established through specific analysis and, in this process, disaster records, history, and experience all play a part. The number of people at risk is also a relevant factor and where relative vulnerability can be assessed (for instance, from possible flood levels relative to population distribution), this is also useful.

14. *Speed of Onset*

As already mentioned, speed of onset is another key evacuation factor. For instance, there is a vast difference:

- between an earthquake and a slowly developing tropical cyclone;
- between a flash flood and a flood which may take days (if not weeks) to build up in a major river system;
- between a controllable narrow-fronted bushfire and a rapidly moving fire which has been whipped up by a sudden, severe wind change.

15. *Warning Time*

Warning time is, of course, closely connected with speed of disaster onset. Obviously, from the examples quoted in the foregoing paragraph, the following alternatives are produced:

- long warning,
- short warning, and
- no warning at all.

These alternatives in warning time are critically important for decision making and for the implementation of evacuation itself.

16. *Methods of Warning*

Methods of warning have significant ramifications for both the decision-making authority and for the threatened community.

The decision-making authority's needs are to ensure:

- First, that the warning gets to all persons to whom it applies;
- Second, that it gets there as quickly as possible;
- Third, that it is given in a clear, understandable, and unambiguous manner; and
- Fourth, that there are alternative methods, in case of failure of disruption.

For the community, it is important to know beforehand how warnings are likely to be transmitted. Also, the community needs to know how it can best "plug into" the warning system.

Methods of issuing warning can vary. They include radio or television broadcast, loud hailer, siren, flags, word of mouth, and so on.

17. *Interpretation of Warning*

Correct interpretation of warning is also critical. For the community, the simple questions are: "What precisely does the warning mean and what are we required to do?" If there are any doubts about the answers to these questions, then the warning system is inaccurate or inadequate, or both.

Doubt in the community's mind about what it should do could have, quite literally, lethal results.

18. *Cancellation of Warning*

There also needs to be some method of cancelling warning, if necessary. Again, the provisions stated in paragraph 17 will apply.

19. *Movement Capability*

Movement capability is yet another key factor for successful evacuation. Two widely differing sets of circumstances illustrate this. In the southern United States which are regularly affected by hurricanes, self-movement is a standard process. Local residents, who usually possess one or two cars, can move themselves and their important personal belongings to safety. All they need to do is to drive north up the nearest convenient highway. By contrast, people living in the southern delta region of Bangladesh have very limited transport resources. They therefore have to rely on government-provided measures (e.g., cyclone shelters) as a way of mitigating the impact of regular cyclone and storm surges. The consistently sad record of casualties shows the limitations involved. The harsh, basic fact here is that if people cannot move, or be moved, there is no point in making an evacuation decision. This may be well a situation in which, under extreme circumstances, disaster managers could find themselves.

20. *Movement Control and Coordination*

The control and coordination of evacuation movement also need careful consideration, both in planning and execution. Some of the key factors are:

- Numbers of people involved;
- Types of transport used(land, air, sea/river);
- Number of vehicles, aircraft, etc;
- Designation of routes, and alternative routes if necessary;
- Necessity for enroute stops;
- Identification of movement coordinating staff;

- Enroute communications; and
- Feeding and refuelling arrangements.

21. *Reception Requirements*

A great deal of care usually needs to be given to arrangements for the reception of evacuees. People who have been uprooted from their own environment, particularly at short notice and in dangerous conditions, are bound to be under some measure of stress. Aspects of reception which need to be covered are:

- General care, welfare accomodation;
- Registration;
- Inquiries (e.g., from relatives);
- Feeding;
- Medical and health arrangements;
- Communication with the evacuation authority;
- Assurance of security for evacuated homes and property; and
- Evacuee self-help, where possible.

Reception is perhaps the most sensitive part of the total evacuation process. It can have many sociological and psychological repercussions; and it usually calls for expert advice on many of its aspects.

22. *Arrangements for Return*

Evacuation plans also need to include arrangements for the return of evacuees to their home areas; or alternatively, in extreme cases, relocation to other areas.

## *Case Study Information*

23. Case studies of various evacuations have produced some interesting slants on the problems involved. Some of these are quoted briefly below.

- *Unnecessary evacuation*
  Several cases have been recorded where precautionary evacuation proved to be unnecessary. Such cases usually produce considerable discontent in both evacuees and recipient communities. In some instances, evacuee groups have defied official regulations and returned to their home areas.

- *Incompatibility*
  Incompatibility of evacuees with their surroundings, and vice versa, has also been a recurring theme. This arises particularly when evacuees are forcibly billetted on families in the recipient areas, or among communities of a different religion or culture.

- *Resistance to relocation*
  Evacuation which develops into permanent relocation can also prove troublesome. The roots of people to tribal or traditional land can go very deep indeed. This has led to rejection of new land allocation and return to former living areas, even though the latter may have been badly affected by salination (from storm surge) or similar disaster causes.

- *Inadequate shelter accommodation*
  There is a good deal of evidence of precautionary evacuations resulting in large numbers of people being sheltered in structurally inadequate accommodation; for instance, churches or community halls. Subsequent buildings collapses, under cyclonic of storm conditions, have caused considerable loss of life.

## Summary

24. The need for evacuation is likely to be a constantly recurring one in disaster circumstances. Whether there is long warning, no warning, or short warning, disaster managers should be able to handle the evacuation process with reasonable effectiveness. However, evacuation efforts are likely to go wrong:

- If there has been insufficient study and analysis of the disaster threat;
- If preparedness levels are inadequate;
- If disaster-prone communities are attuned to the possibilities (and the risks) of evacuation; and
- If everyone, including the community, is not absolutely clear about the responsibility for decision making and the evacuation process.

The problems of evacuation are clear. So are the various factors involved. They have been spelled out in practical terms over many years. Therefore, if the requirements are followed, there seems very little reason why evacuation failure should occur.

*Chapter 18*

# Logistics

## Purpose

1. The purpose of this chapter is to outline some major factors which apply to logistics in disaster management. Logistics constitute a primary component of response operations (which are discussed in Chapter 17).

The chapter includes:

- A general description of logistics;
- The importance and place of logistics in disaster management;
- Response aspects;
- Recovery aspects;
- Preparedness requirements; and
- Some management, administrative, and accounting considerations.

2. The chapter does not cover the logistic aspects of major refugee operations of famine relief. These can be found in the Handbook for Emergencies issued by the United Nations High Commission for Refugees (UNHCR).

## General Description of Logistics

3. Logistics have been described as the procurement and delivery of

- the right supplies,
- in the right quantities,
- in the right order,
- in good condition,
- at the right place, and
- at the right time.

It is interesting to note that before the word logistics came into wide international usage (principally in the military field during World War II) an alternative term was sometimes "supplies and transport." These two latter activities, in fact, constitute the major components of logistics as currently applied to disaster management. They are certainly the two activities which are likely to be of most direct and practical concern to the majority of disaster managers.

## The Importance and Place of Logistics in Disaster Management

4. In line with the foregoing description and as stated in paragraph 1, logistics are of primary importance during response operations. However, they also play a key role in recovery programs (see Chapter 19). Consequently, logistic considerations need to be taken fully into account during preparedness assessments.

## Response Aspects

5. In the context of response operations, logistics are especially important in three interrelated areas. These are outlined below and one particular characteristic concerning them needs to be emphasized. This is the urgency which applies to them all. This indicates that areas such as prior planning and organization must receive due attention and underlines the importance of preparedness measures generally.

### 6. *Fulfillment of Operational Tasks*

Operational response tasks usually require a considerable amount of movement by land, sea, inland waterways, and air; and action by counter-disaster units and teams, much of which may be carried out on a deployed task-force basis.

These response tasks require the use of various forms of operational capability, usually in the form of vehicles, boats, and aircraft. They also require the availability and use of various forms of supplies. Such supplies may include:

- Petrol, oil, and lubricants;
- Technical spare and repair parts (depending on circumstances, these may have to cover a wide range of items to cope with vehicles, aircraft, communications systems, and similar requirements);
- Personnel subsistence and support commodities, including food, medical and health items, tents, and so on; and

- Administrative items of various kinds; for instance, standard report forms and other requirements for survey and assessment if this type of operational activity is likely to be involved.

## 7. *Procurement and Distribution of Relief Commodities*

The procurement and distribution of relief commodities may involve heavy demands on commodities and transport, as illustrated below.

*Commodities*
In severe disaster conditions, demands on relief commodities are likely to be very heavy indeed. However, in all circumstances, it is advisable to restrict distribution to a reasonable minimum. This is prudent in the community. It is therefore appropriate to restrict items to the well-recognized essentials of food, clothing, shelter, and medical necessities. This type of frugal management approach also helps achieve a streamlined, and thus more effective, logistic operation. Nonetheless, providing essential commodities is not likely to be problem free.

In small countries especially, preparedness stockpiling (even if it is feasible at all) is unlikely to meet operational needs fully. Emergency procurement is therefore likely to be a part of the overall logistics role. Indeed, this is likely to apply to many larger countries as well. One comment on procurement which often arises (particularly in post-disaster review) is that not enough use is made of in-country sources of supply. This largely arises because, under emergency conditions, it often seems easier and quicker to accept and utilize international input. This point is one which needs to be resolved, at least as far as possible, during preparedness action. Thus, as a preparedness measure, sources for emergency procurement need to be earmarked by disaster management authorities, usually from within the local wholesale/retail food chain; local and export manufacturing capability (e.g., for rice, tinned fish, biscuits); special agricultural sources (such as possible emergency harvesting of certain crops); and government stockpiles of items such as medical supplies.

In the context of response operations, however, it needs to be borne in mind that loss of commodities may occur because of disaster impact.

Disaster management authorities therefore need to keep an ongoing operational check of the balance between:

- assessment of commodity needs (as progressively updated by survey and assessment action),

- availability and timing of local (in-country) commodities, and
- availability and timing of international assistance supplies.

*Transport and Transport Systems*

The availability of transport and the serviceability of transport systems obviously constitute a key factor in the effective distribution of relief commodities. Some considerations which are likely to apply include:

- Loss and/or damage inflicted by the disaster;
- Flexibility in transport capacity and systems, especially the ability to switch resources from unaffected areas to disaster-stricken areas;
- Ability to procure transport resources by requisitioning and/or charter;
- Difficulties of transport access to some stricken areas because of remoteness, severed communications, or various severe disaster effects;
- Limitations or benefits resulting from preparedness; for instance, earmarking (or otherwise) of emergency transport capability in plans and departmental standard operating procedures;
- Types of transport available (for instance, remote mountain areas, or areas which are isolated by disaster effects, cannot be supplied unless airlift/airdrop capability is available; limited four-wheel drive capability may also be a restrictive factor); and
- Assistance from international resources, which can augment in-country transport capability or temporarily make good unserviceable portions of such capability.

It is noteworthy that, in many disaster situations, the withdrawal (for whatever reasons) of international transport assistance (e.g., helicopters or aircraft capable of operating off short airstrip) nearly always causes a marked decline in supply capability, albeit for a temporary period only.

A major consideration which may help disaster managers in the transportation of supplies is the optimum use of self-help from stricken communities. For instance, in many cases it has been found expedient to deliver bulk supplies (by boat or airdrop) to various convenient points from which they can then be collected by responsible community representatives or groups. This presupposes, of course, that:

- adequate communication can be established between the disaster management authority and the stricken communities; also that

- such communities still remain a basis of local government or traditional leadership and authority.

## 8. *International Assistance Activity*

From a logistics viewpoint, international assistance activities involve important factors. For example:

- International assistance usually boosts the availability of much-needed relief commodities, provided liaison between the stricken country and international funding agencies has ensured the preclusion of unnecessary relief items.
- If, however, good liaison is not maintained, inappropriate and often useless items may be sent. This can be a serious liability, since the in-country supply system may become choked and valuable local resources may have to be developed to sort usable commodities from recipients to ensure that inappropriate supplies (e.g., unacceptable foodstuffs) are made known to funding agencies.
- International assistance can provide additional transport capability, especially of the emergency kind such as helicopter lift and four-wheel drive vehicles.
- International assistance may also be invaluable in the handling and distribution of supplies.
- International relief input usually places additional demands on the in-country logistic system. This may be a crucial factor if major ports, airfields, roads and railways have had their capacity reduced by disaster effects. Extra demands may also be placed on fuel and food stocks by visiting aircraft and various relief teams.
- In the interests of both the stricken country and the international assistance agency, it is important that no undue delays are imposed on international inputs by delays from customs or similar formalities.

From the foregoing points, it is clear that international assistance activities, while contributing many invaluable benefits, may also impose logistic complications. Any such complications need to be minimized through prior planning and preparedness arrangements.

## 9. *General*

On response aspects, heavy demands can invariably be expected on:

- movements capability, provided from both in-country and international sources;
- commodities required to provide relief for stricken communities; and
- commodities necessary to support response operations themselves.

Moreover, because of

- possible loss and/or downgrading of logistic capability (i.e., transport and supplies) due to disaster effects;
- urgency in response operations, especially to relieve hardship and suffering; and
- complications of integrating international assistance effort,

disaster managers are likely to be confronted with major problems of decision making, allocation of priorities, and coordination of logistics operations.

## Recovery Aspects

10. Recovery programs, like response activities, have considerable logistics implications. In fact, if such programs are to be accurately identified and defined, logistic capabilities and limitations must be taken into account throughout the total process of:

- Producing an adequate database from which to define various aspects of the recovery process (for instance, this database needs to include a general assessment, at least, of logistic requirements and available capability);
- Defining the overall strategy for recovery and associated facets of national development;
- Determining the individual recovery programs within the overall strategy; and
- Implementing individual programs and projects on a well-managed and coordinated basis.

### 11. *Recovery Program Assets*

From a logistics angle, the recovery process offers certain assets when compared with response. For instance:

- Logistics aspects can be integrated into recovery programs with considerable certainty and reliability.
- Recovery programs and projects take place in more stable conditions.

- Less pressure and urgency are likely with regard to decision making and the implementation of projects.
- There is likely to be more time available for the management of commodities and supplies, the maintenance of transport and transport systems, and the conservation of personnel resources.

### 12. *Recovery Program Liabilities*

Conversely, recovery programs carry certain liabilities. These may include:

- A general downgrading on in-country logistics capability, caused by the heavy demands of response operations. This downgrading may be exacerbated by the post-disaster condition of roads and railways. Such downgrading may lead to a conflict of priorities between the need to rehabilitate roads and railways, and the urgency of recovery projects.
- The withdrawal or reduction of international assistance capability.
- The heavier logistics lift demands of recovery programs (for example, the movement of very large components such as bridge structures).

### 13. *General*

From a general recovery standpoint, logistics considerations are critically important. They have the same influence on programs and projects as, for example, availability of supplies and workforce capability. They therefore need to be reviewed as an integral component of such programs and projects.

## Preparedness Requirements

14. It is evident from what has been said in paragraphs 5–13 above that considerable preparedness action has to be taken if the logistics demands of response and recovery are to be satisfactorily met.

### 15. *Planning and Organization*

First, there must be an adequate basis of planning and organization. Logistics can be a brittle component in response and recovery, particularly the former. This is because of the variable conditions and requirements which may apply. Therefore, anything that can be done through effective planning and organization to clarify, facilitate, and implement logistics action is likely to be invaluable. Also, given that logistics have to cover a wide range of possibilities, good planning, and organization can help minimize the risks of gaps and duplication.

## 16. *General Assessment of Logistics Capability*

Within the foregoing general framework of planning and organization, it is then necessary to make a general assessment, in the preparedness sense, of logistic capability. This assessment should primarily take into account:

- Potential disaster scenarios and logistic needs;
- Vulnerability of logistic components (i.e., commodities, transport vehicles of various kinds, supply, and routes).

Such an assessment could be expected to give a reasonably accurate picture of how logistic aspects would apply to a particular disaster situation. For instance, in the case of major flooding, a forecast could be made of:

- what supplies and transport facilities might be lost or cut off, and for what period (e.g., supplies inundated, roads cut or washed away);
- what areas would be accessible or inaccessible for supply purposes;
- what effects the evacuation of communities would have on logistic requirements, in terms of both evacuation movement and the subsequent supply of essential commodities to the evacuees;
- what special logistic items might need to be brought in to reinforce existing capability (e.g., special flood boats for rescue and/or relief purposes);
- what preparedness and associated arrangements could be taken to safeguard logistic component, either by preventive arrangements (e.g., sand bagging) and/or relocation.

## 17. *Cataloging of Logistic Tools*

In conjunction with this suggested assessment, it is also necessary to categorize and record what might be termed the logistic tools to meet disaster situations. Examples of such major categories are given below:

*Commodities likely to be needed*

- Operational support items (e.g., fuel);
- Relief commodities (e.g., food, shelter, materials, clothing, medical necessities, water purification accessories); and
- Items likely to be required for recovery programs (e.g., building materials).

*Sources of supply*

In paragraph 7 above, the following possible sources of supply were suggested for emergency procurement:

- the local food chain,
- local manufacturing sources,
- special agricultural sources, and
- government stockpiles.

Other sources of supply include:

- Red Cross/Red Crescent which, in a number of countries, holds special disaster-related items, such as medical supplies, cooking utensils, tools, and special family kits.
- NGOs, some of which may have special capabilities, such as the provision and erection of large temporary accommodation units;
- International funding agencies, including service-type organizations (Lions, Rotary, Apex, etc.), which will undertake self-contained recovery projects, such as the rebuilding of schools or medical clinics.

Obviously, individual circumstances will differ with regard to sources of supply. Thus, disaster managers will need to adapt their own detailed approach from the above general suggestions. It is also worth noting that some supply problems can include the donation of:

- inappropriate items of food and clothing;
- soiled, unsuitable items of various kinds;
- food which is unfit for consumption; and
- out-of-date medicines and drugs.

(See also paragraph 8 above).

*Storage*

Storage requirements may include:

- Main bulk storage of a permanent nature (e.g., large government or commercial stores or warehouses);
- Bulk storage at major distribution points (e.g., in or adjacent to stricken areas). Such storage may need to be weatherproof and secure, especially if perishable and valuable items are involved.

Requisitioning of buildings may therefore be necessary or alternatively.

- Temporary storage (e.g., in port and airfield areas) for stocks in transit;
- Temporary holding storage at forward distribution sites (e.g., advanced airstrips).

*Handling*

Handling requirements at entry, storage, and transit points are also an important consideration. Needs may include forklifts, trolleys, cranes, and similar items. It is also important to ensure (by prearrangement if possible) that where manhandling is anticipated, packages can be handled by one person; failure or inability to take this kind of precaution can result in very wasteful use of available personnel.

*Sources of transport*

Possible sources of transport which need to be assessed on preparedness basis are:

- *Road*
  Government transport capability;
  Commercial transport, for charter or requisitioning;
  Private vehicles, for charter or requisitioning; and
  International agencies and contractors working on in-country projects.
- *Rail*
  Government rail system, with assessment of haulage/movement capability.
  Commercial light rail systems, if available, such as special industrial/commercial links to and from ports (for sugarcane, woodcutting, or similar projects).
- *Inland waterways*
  River and canal shipping systems, with capabilities and restrictions.
  Government and commercial craft;
  Private craft, including local boats and canoes, which may provide vital transport to remote locations.
- *Sea, coastal, and interisland shipping*
  Government vessels;
  Commercial vessels, for charter or requisitioning; and
  Private craft (e.g., launches and motorboats).
- *Air transport* (fixed wing and helicopters)
  Government airlines or aircraft;
  Commercial airlines and operators; and
  Private aircraft.

In the above resources of transport, it is assumed that government categories would include military resources, as appropriate.

*Factors affecting movement*
Factors which may affect movement are:

- Availability of maps, charts, navigation aids, timetables, and similar information;
- Terminal and transit factors, such as port capacities/facilities, airfield capacities, road states (including weather effects), bridge weight capacities, and so on.

## Management, Administrative, and Accounting Considerations

18. It is evident from the foregoing that logistic activity is bound to involve management, administrative, and accounting considerations. These are summarized below.

### 19. *Management considerations*

Management must be primarily designed to achieve effective direction and coordination of logistic effort. This can be conveniently divided into operational and routine considerations.

- *Operational considerations*

  *In response operations*, it is widely accepted that logistic aspects can be effectively integrated through the EOC (emergency operations center) system. As an example, the logistics desk in one national emergency operations center was given the following responsibilities:
  - To receive, store, secure, transport, and distribute relief supplies.
  - To coordinate supplies distributed directly by NGOs and other organizations.
  - To ensure proper maintenance of vehicles and equipment.

  In another case, logistic requirements were dealt with by:
  - A supply and finance desk, plus
  - Subdesks for road transport, air transport, and shipping.

  Both these options worked satisfactorily and that disaster managers should use whatever detailed system is appropriate and compatible at all applicable levels of government (e.g., national, provincial, and local).

  *For recovery programs*, it is usually possible to take a somewhat more general management approach. One example comprised:

- A national Office of Relief and Reconstruction, the roles of which included overall logistics management, plus
- The covering of provincial logistic requirements through normal government processes.

A second example consisted of:
- A national executive committee at the ministerial level;
- A support structure from the NDC and a special technical advisory team, which covered logistic requirement on behalf of the national executive committee, plus the covering of provincial logistic requirement through normal government process.

- *Routine considerations*

Routine logistic considerations (i.e., during nondisaster times) are normally the responsibility of individual government departments and NGOs. Counter-disaster plans, and more particularly standard operating procedures, should provide necessary guidelines covering both supplies and transport. (See also Chapter 13 on use of resources.) Routine management is also materially assisted if a national disaster management office (or the equivalent) is given responsibility for assisting in the planning, organizational, and preparedness measures which relate to logistics.

## 20. *Administration and Accounting*

Logistic aspects usually involve heavy demands on administrative and accounting systems, mainly because of:

- procurement of commodities,
- charter and requisitioning of transport and other facilities,
- stock control during urgent relief operations,
- recovery of equipment and tools issued for emergency operations,
- handling of international commodity inputs
- need to account accurately for in-country and international relief fund donations,
- special demands on national expenditure, and
- other similar requirements.

It may therefore be advisable for governments to consider some or all of the following arrangements:

- If counter-disaster legislation is implemented, it should cover the release of appropriate funds from national government resources.
- In allocating responsibility for disaster management at the national and provincial levels (i.e., to ministers or provincial governors), the national government should specify what delegation of expenditure applies, or, alternatively, how urgent expenditure needs are to be met.
- It may also be advisable to set up a special system for administration and accounting to cover disaster requirements. Such a system can be kept simple by using number-coded or color-coded forms.
- Special arrangements may be required for the urgent release or issue of government goods and equipment.
- A special department or section may have to be set up to finalize administrative and financial matters, as well as to assess the cost of the total counter-disaster effort. This may be particularly advisable if any attempt is to be made to calculate cost–benefit with regard to future mitigation programs.
- Arrangements may be possible with international assistance agencies to obtain assistance in offsetting disaster costs.
- It may be possible to arrange for the employment of temporary or auxiliary personnel (e.g., former government employees) to assist in dealing with administrative overloads arising from disaster.

# PART V

# Major Post-Impact Factors

## Chapter 19

# Recovery

## Purpose

1. For the purpose of this handbook, the following definition is applied to recovery:

> Recovery is the process by which communities and the nation assisted in returning to their proper level of functioning following a disaster.

The recovery process can be very protracted, possibly taking 5–10 years or even more. Recovery is usually taken as including other aspects such as restoration and reconstruction.

2. The purpose of this chapter is to outline the major factors which apply to recovery, with particular reference to the following:

- Key points from disaster analysis,
- Transfer of responsibility from response to recovery,
- Continuation of response activity,
- The basis of recovery action,
- Problem areas in recovery,
- Major requirements for effective recovery,
- Human factors in recovery,
- Resources relevant to recovery programs, and
- Projection of the recovery process.

## Reference to Other Aspects of Recovery

3. Various references to recovery have been made in preceding chapters of this handbook. The main such references are given below.

- Chapter 3 contains various references to the place of recovery in national disaster management policy.

- Chapter 5 refers to the place of recovery in the disaster management cycle.
- Chapter 6 contains reference to the linking of disaster management activities, including recovery, with national development planning.
- Chapters 8 and 9 contain references to the use of international resources in recovery programs.
- Chapter 10 discusses the ramifications of political leadership on the recovery process.
- Chapter 11 outlines various organizational aspects which apply to recovery.
- Chapter 12 refers to the need to include mention of recovery policy in national disaster plans.
- Chapter 17 discusses the follow-on activity from response operations into the recovery process.
- Chapter 18 contains reference to the logistic aspects of recovery.

## Key Points from Disaster Analysis

4. Analysis of disaster events has identified various patterns and timescales for recovery activity. It also identifies the following key points for disaster managers:

- In most sizable disasters, there is no quick finalization of response/ recovery action.
- The potentially long recovery period means that the vulnerability of a stricken country to future disasters is exacerbated.
- The financial and material demands of recovery programs usually restrict resources which might otherwise be available for strengthening other management measures (e.g., mitigation and preparedness).

The cumulative effect of the above factors is often clearly seen in very disaster-prone countries. In such countries, the repeated impact of disaster means that recovery is never really complete and that preparedness remains restricted. Thus, the countries seldom reach a stage when they can cope satisfactorily with their disaster threats.

## Transfer of Responsibility from Response to Recovery

5. In Chapter 17 on response, reference was made to some problems which affect the transition from response activities to recovery programs. It was indicated that there tends in fact to be a somewhat blurred period between such activities and programs.

6. An important factor in this transition process is the actual transfer of authority. The transfer is, in fact, one which perforce takes place between two dissimilar processes; one (response) depends on urgency and short-term expediency, while the other (recovery) has to be of a more general, considered, and long-term nature. To amplify this point, the following considerations are relevant:

- Response action is usually dealing with urgent problems which will not wait; they will only get worse unless urgent solutions are applied.
- Response action often takes place under extraordinary powers (e.g., a state of emergency or a declared national disaster). It can therefore go beyond normal day-to-day limits of authority. Such freedom of action does not apply to recovery.
- Response action must sometimes be taken without consideration of longer-term repercussions (e.g., effects on recovery programs).
- Some response actions may have direct implications for recovery programs (e.g., large-scale evacuation movement, the need to accept outside assistance which upsets the balance of the local economy).
- Response action may have lowered the capability of key government departments (by using up financial allotments, fuel allocations, etc., and by lowering vehicle serviceability). Thus, when these departments are called upon to participate in recovery programs, this necessarily downgrades overall recovery capability.
- Recovery action may also have to operate within a changed community attitude. Emergency response action is usually accepted and supported by stricken communities because, basically, it is relieving their immediate problem(s). However, when the urgency of such problems has passed, communities tend to be more questioning, and perhaps demanding, concerning action being taken on their behalf.

In sum, therefore, when responsibility is transferred from response to recovery agencies, the latter are not only working within a blurred period; they may also face carryover problems from response and, moreover, be forced to cope with such problems with restricted capability.

## Continuation of Response Activity

7. A further factor in recovery is the need to deal satisfactorily with response activities which continue, or have to be projected, beyond the end of the emergency response period (e.g., continued welfare programs). From an overall recovery viewpoint, these continuation activities need to be expeditiously dealt with on two counts:

- First, as programs in their own right, in the public and national interest;
- Second, as part of the total strands of relief action which need to be carried through from response operations into longer-term recovery programs. For instance, short-term emergency feeding programs may have to be extended into special school nutrition programs, and temporary shelter arrangements may have to be converted into special rehousing projects. (See also Chapter 17.)

## The Basis for Recovery Action

8. Given that the two foregoing issues of responsibility transfer and continued response activity are recognized and suitably dealt with, the basis for recovery action rests primarily on the following four pillars:

- Production of an adequate information base from which to define the various necessary aspects of the recovery process;
- Definition of an overall strategy for recovery, compatible with proposed and potential national development;
- Determination of individual recovery programs within the overall strategy;
- Implementation of individual programs within the overall strategy; and
- Implementation of individual programs and projects on a well-managed and coordinated basis.

These aspects are amplified below.

## Problem Areas in Recovery

9. To put recovery into its full perspective, it is necessary to outline some of the typical problems which can arise, namely:

- Delays occur in the formulation of recovery programs because the definition and management of recovery programs have not been adequately considered in overall counter-disaster planning and organization.
- Damage and destruction may be so severe and extensive that it is difficult, and takes considerable time, to assess and formulate recovery programs.
- Information is inadequate for the formulation of recovery programs. In some cases, this can necessitate resurveys in order to establish post-disaster effects more accurately.

- The additional load imposed on the government system by recovery programs may slow down the functioning and output of government departments, so that the whole recovery process becomes unsatisfactory.
- Occurrence of another major disaster (not necessarily of the same kind or in the same area) may divert funds and resources from recovery programs.
- Problems relating to ministerial responsibilities may arise because recovery requirements overlap from one department to another.
- Restrictions on availability of finance (and/or uncertainty concerning international assistance funding) may hinder the formulation of recovery programs.
- Political problems can sometimes arise if certain areas/communities appear to not be receiving the same priority of attention as others in regard to recovery programs.
- Inadequate attention may be given by governments to public information programs and public relations generally, so that false impressions or perceptions arise concerning recovery measures.
- Some major and vital reconstruction programs may take a long time (e.g., reconstruction of bridges, roads, and rail systems). This may cause economic loss, further hardship, and other problems for disaster-affected communities.
- Limitations may apply to the implementation of recovery programs; for instance, in materials, equipment, specialist and skilled personnel.
- The time process of some recovery programs cannot be hastened; for instance, those which apply to the natural recovery of, say, coconut palms or oil palms. This usually has economic implications.
- Some recovery programs may necessitate substantial reorientation in national activities. For instance, disaster may affect an export market commodity (such as agricultural produce). The vacant market may then be taken over by another country and may be lost to the disaster-affected country for the foreseeable future.

The above items do not necessarily reflect all the problems which might affect recovery programs. But they illustrate the kind of difficulties with which disaster management officials may have to deal.

## Major Requirements for Effective Recovery

10. With the above possible problem areas in mind, major requirements for effective recovery are outlined below. In most cases, it is likely that, before recovery action can be established, discussion with international assistance agencies will need to take place. If (as suggested at various places in this

handbook) routine national/international disaster management contacts are maintained, the international assistance aspects of recovery time may have to be given over to assistance meetings and negotiations. In any event, before recovery action can be firmly established, the stricken country needs to have a reasonably clear idea of what outside assistance it can expect.

## 11. *Understanding the Recovery Process*

Ministers, their advisors, and others concerned with the formulation and direction of recovery programs should fully understand reconstruction and long-term community rehabilitation. The recovery process is therefore complex and extensive; it can generate a multiplicity of problems, some of which will necessitate changes to original plans as recovery proceeds. Thus, senior decision makers should be prepared to take a flexible attitude toward policy implementation to produce the best results. Also, it is important to bear in mind that recovery from disaster offers opportunities for national improvement (sometimes called the "disaster as a benefit syndrome").

## 12. *Recovery and National Development*

The need to implement disaster recovery programs often has direct implications for national development plans. Where possible, therefore, it is desirable to reconcile appropriate aspects of recovery with national development planning. This has particular relevance to major restoration and reconstruction projects.

## 13. *The Recovery Information Base*

As already indicated, a primary prerequisite for recovery action is the availability of adequate and relevant information. Major sources for such information are as follows.

- *Information from response operations*
  This may include:
  - information from damage surveys and needs assessments;
  - various forms of operational reports;
  - departmental and other reports on completion of emergency phase; also, similar reports from NGOs;
  - information collected by emergency operations centers;
  - reports from international assistance agencies;

– media information; and
– various submissions by individuals.

It is important that this information is in a form which can be used for recovery purposes. This may necessitate special processing, editing, or other similar treatment.

- *Post-disaster review*
  If post-disaster review can be carried out reasonably soon after completion of the emergency phase, the information acquired can also be utilized for recovery program purposes. For instance, this information may indicate that certain land-use restrictions need to be applied in the future. These, in turn, could have repercussions on longer-term recovery projects.

- *Information from development programs*
  As indicated in paragraph 12 above, recovery requirements may have a direct bearing on existing development plans (e.g., the development of major road systems). Information concerning such development must therefore be taken into account for recovery purposes.

- *Information from special teams*
  If a special technical advisory team is used to monitor recovery needs from the immediate post-impact stage, the information it can provide is likely to be of major value and importance.

- *Information for program parameters*
  Information which affects the size and shape of recovery programs also needs to be taken into account. For instance:
  – project costs,
  – timescales,
  – in-country input capability, including financial, material, and human resources,
  – international assistance potential.

- Information from previous disasters
  Documented information from previous disasters, especially concerning recovery programs, can also be of considerable assistance.

## 14. *Overall Recovery Strategy*

The determination of overall recovery strategy obviously needs to be made at the top government level. Exactly, how this is done must depend on individual circumstances. However, in this regard, the advantages for having a small technical advisory team to monitor post-disaster requirements from impact onwards, can be substantial (see also Chapter 17 on response). The strategy, when finalized, needs to be broad in nature and to include, for example:

- major areas of recovery need,
- envisaged time frame for recovery action,
- interlinking of recovery with national development (see paragraph 12 above),
- broad assessment of resources available, and
- system for monitoring recovery from national government level.

## 15. *Policy Direction of Recovery Programs*

Following from the overall strategy, it is important to ensure that policy direction for recovery programs is clearly established and defined. Several options for the direction of programs include:

- A special ministerial committee;
- One minister, with a special temporary office or section, acting on Cabinet direction;
- Cabinet itself, providing direction to individual ministers and their departments.

Whatever system is used, the important need is that there should be no doubt where responsibility lies for the overall direction of recovery action.

## 16. *Determination of Program Needs and Areas*

Analysis and assessment of all relevant information will enable individual recovery programs and the areas they are required to cover to be identified. Such programs are likely to include:

- Government aspects
  - National infrastructure (roads, ports, railways, etc.);
  - Government administrative facilities;
  - Education facilities;
  - Health-care systems; and
  - Resettlement of displaced persons and communities.

- Private sector
  - Industrial systems.
  - Commercial buildings, stores, shops, tourist facilities, etc.

- Community
  - Reestablishment of social services system.
  - Long-term rehabilitation of communities and individuals.

## 17. *Assessment of Resources*

Most recovery programs, especially those concerned with major reconstruction, tend to be costly. Early in the recovery process, it is important to establish what finance is available to carry out these programs. Similarly, an assessment of resources (such as equipment, materials, skilled personnel) needs to be made. These aspects are particularly important in determining the size and shape of the total recovery program.

## 18. *International Disaster Assistance*

International disaster assistance is usually generous, especially in the case of severe disasters. As indicated in Chapter 17 on response, such assistance plays an important part in emergency relief programs. It is also invaluable in recovery programs, where it usually takes the following forms:

- direct financial donations, grants, or credits;
- equipment and materials;
- project implementation (e.g., an overseas country will undertake a school rebuilding program or restoration of a wharf);
- international welfare agencies or religious bodies will help their counterparts in the stricken country.

## 19. *Program Decisions and Priorities*

As early as possible in the recovery process, firm decisions need to be taken on:

- the types of programs to be implemented, and
- the priorities which should apply.

These decisions must obviously be taken at the national government level, especially in view of their possible connection with national development planning. When these program decisions and priorities are being made, it may also be worth considering what emergency action should apply to them if disaster strikes again, before they are completed. This could be particularly relevant in disaster-prone countries.

## 20. *Program Management and Monitoring*

Effective management and progress monitoring of recovery programs is essential, particularly if such programs are complex and numerous. This is where the special minister/office system (see paragraph 15) has some advantages, because it concentrates action and detailed decision making into one place in government. Also, it provides an easily recognizable focal point for inquiries and information; this is especially desirable if the follow-on programs from response are extensive. However, in many cases, governments have chosen to follow the principle of returning to the normal systems and processes of government as soon as possible. This has usually been done on the basis that such systems and processes are well-established and understood; thus they are likely to be more effective than some specially introduced alternative. In these cases, programs have usually been directed from Cabinet, through individual ministerial responsibilities.

## 21. *Continuation of Emergency Relief Programs*

It is important that programs running over from the emergency phase (e.g., emergency feeding programs) should continue to receive the necessary priority. If not, they can easily become overshadowed by the larger recovery programs. Such follow-on programs therefore need to be monitored and assessed alongside recovery programs, since they are part of the total recovery process.

## 22. *Disaster Management Aspects*

Since recovery programs tend to be managed through normal government processes, participation in them by disaster management authorities may be fairly limited. However, disaster management aspects should not be overlooked, because major reconstruction projects, if not properly evaluated, may produce disaster-related problems in the future. To avoid such possibilities, there should be an ongoing system of liaison between national development planning and disaster management.

## 23. *Public and Media Information*

It is obviously desirable that the public and the media should be kept appropriately informed of various aspects of recovery programs. Special arrangements may therefore be necessary to provide briefings and other dissemination of information. The special minister/office system for recovery program management is, of course, well suited for fulfilling this information requirement.

## Human Factors in Recovery

24. A great deal of material has been published concerning the physiological, psychological, and sociological effects of disaster, and this is widely available to disaster managers who wish to study these aspects in detail. For the purpose of this chapter, however, the inclusion of certain practical references is considered appropriate, as below.

### 25. *Disaster Victims' Attitudes*

An exceptionally experienced disaster relief official has noted, after no less than 44 years of direct involvement with disaster victims, that five post-disaster characteristics become clearly discernible.[1] These are summarized below.

(1) *Devastation*
This stage manifests with general shock and a certain numbness or failure to comprehend the full implication of the situation. The victim is prone to burst into tears and repeat over and over again certain aspects of the damage which have indelibly imprinted themselves on his/her memory. Sometimes victims almost appear hard of hearing or mentally incapable of comprehending what one is saying.

(2) *Realization*
The realization period appears to develop with particular emphasis when the person returns to the site of their now destroyed dwelling. They search amid the rubble for specific items, such as memorabilia, old rings and brooches, something that belonged to mother or grandfather, and so on. Gradually an overwhelming depression comes with the realization that all things, or at least most of them, that made up their background security have disappeared.

(3) *Rationalization*
After a few days, the victim starts to assess his/her past, present, and future.

First, the question that comes to a victim's mind is "Why did this happen to me?" It has been surprising to learn from a number of persons that their minds run along similar tracks. Many people seem to develop the idea, particularly elderly people, that God is punishing them.

---

[1] Dame Phyllis Frost, Chair, State Relief Committee, Victoria, Australia, in a report on a major bushfire disaster in 1983.

The rationalization of the present takes most people in a constructive way and they start immediately cleaning up the area that has been left, making provision for the children's schooling, incorporating the help of relatives and friends, and gratefully receiving any government or private aid that is forthcoming.

The rationalization for the future consists of assessing what will be best for the family, now that the previous dwelling and security items have been lost.

(4) *Accusation*
After the victims have sorted themselves out, they seek, through the mass media or in other ways such as public meetings, to accuse those people whom they consider have been responsible for their suffering. This has increased with the advent of television and increased new facilities through the mass media and is encouraged by those involved in preparing and presenting programs through the mass media.

(5) *Accumulation*
When the victim has developed to the fourth stage, the fifth phase starts creeping in; this is the development of greed and avarice to such an extent that many victims take the emphasis off what they really need to be settled back into their homes, and start thinking in terms of what they would really like or want, which is a great deal more than their needs. This is the most unsavory factor that I have found when dealing with victims.

There will almost certainly be variations of these characteristics in different countries and circumstances. However, the ones quoted serve as a good example of the frame of mind of victims when they become involved in the recovery process.

## 26. *Relocation of Communities*

In a number of recovery circumstances, it has been found necessary to relocate communities or groups of people; for instance, in circumstances such as:

- complete loss of traditional subsistence land and dwellings because of major earthquake/landslide/riverbank erosion;
- repeated damage to or destruction of living areas by persistent flooding.

In general, under such circumstances, relocation results in a number of problems, even though the new environments tend to be better and safer to live in. Most of these problems tend to arise from human factors (e.g., being torn away from traditional roots) rather than from material ones. Predicted rises in sea levels may cause major relocation problems for some countries in the future.

## 27. *Evacuation*

Human factors can also influence evacuation. While evacuation is usually a response measure, at least in the first place, its effects often spill over into the disaster management aspects of recovery. This particularly applies if evacuees cannot be returned to their home areas reasonably quickly. In such cases, the evacuees themselves can become restless and worried by their enforced absence from their own land and property; also, the longer they stay in the evacuation zone, the greater becomes the possibility of friction with the local population.

## 28. *Resettlement*

Even if relocation, evacuation, and similar problems are confined to a minimum, human factors are still likely to bear on the general resettlement of communities into a normal pattern of life. In the recovery period, people can often be affected for considerable periods by

- personal injuries and trauma,
- family bereavement,
- loss of property and possessions,
- damage to dwellings, and
- reduced means of livelihood.

These factors tend to generate problems which can affect an already overburdened recovery process. For instance, they can lead to reduced efficiency in people who are required to implement recovery programs, and increased load on social security and similar agencies.

## Resources Relevant to Recovery Programs

29. The resources most relevant to recovery programs can be summarized as follows:

*Resource organizations responsible for providing data for, and assessment of recovery programs*

- disaster management authorities;
- government planning authorities;
- government departments, organizations, and agencies;
- international funding agencies;
- NGOs; and
- statistical, information, and media agencies.

*Resource organizations responsible for deciding on types and numbers of recovery programs and relevant priorities*

- national government, assisted as appropriate by regional and local governments and NGOs (plus supporting advice from international funding agencies).

*Resource organizations responsible for directing and monitoring recovery programs*

- minister and departments, and/or other authorities, as designated by national government.

*Resource organizations responsible for implementing recovery programs.*

- government departments, organizations, and agencies;
- private sector contractors and other resources;
- international assistance teams;
- international and national welfare organizations; and
- program specialists.

*Organizations capable of providing support for implementation of recovery programs*

- NGOs;
- community groups, including religious bodies;
- other volunteers; and
- media and information agencies.

## Projection of the Recovery Process

30. In the total process of recovery from disaster, nations hope to achieve maximum possible benefit from possible benefit, especially in terms of enhanced infrastructure, counter-disaster capability, and general development. Unfortunately recovery, particularly in its early stages, is usually a period in which government capability is still downgraded because of disaster impact. At the same time, political pressures may arise because the disaster-affected population becomes impatient at delays in the restoration of its former amenities and subsistence level. All this can and often does lead to a situation wherein the recovery process is implemented hurriedly and is not therefore projected deeply enough into the long-term future.

31. Therefore, in the formulation of recovery programs, it should be recognized that, in the disaster cycle, the four segments which follow recovery are development, prevention, mitigation, and preparedness, in that order. If these four segments or components are visualized as providing a total combined field for post-disaster exploitation, then recovery takes on a very far-reaching potential. The following example illustrates this point.

*Restoration of a coastal road system following damage caused by storm surge/ severe wave action can:*

- improve transport movement for business, tourism, social, and disaster response purposes;
- through the inclusion of seawall protection measures, help mitigate the effects of future disasters;
- through careful rerouting, place reconstructed en-route human settlements further inland, thus reducing future community risks;
- improve drainage so that the road system is less vulnerable to the effects of flooding.

This multipurpose approach to recovery programs is worthy of attention from the viewpoints of government policy, national development, and disaster mitigation/management.

# Chapter 20

# Post-Disaster Review

## Purpose

1. The purpose of this chapter is to outline the major considerations which apply to post-disaster review.

2. The chapter covers:

- the importance of post-disaster review,
- some reasons for the omission of such review,
- aspects covered by review,
- major investigations into disaster, and
- use of results.

## Other References to Post-Disaster Review

3. Other references in this handbook to post-disaster review are:

- Chapter 3 refers to the use of post-disaster review as a means of helping check the validity of national disaster management policy.
- Chapter 5 indicates the place of post-disaster review in the disaster management cycle.
- Examples of disaster legislation contained in Chapter 7 include the need for an annual report on disaster-related activities. Such reports could constitute a form of review.
- Chapter 12 contains various references to the place of post-disaster review in plans.
- Chapter 14 emphasizes the importance of post-disaster review in the monitoring of preventive measures.
- Chapter 16 quotes post-disaster review as an element in assisting to maintain levels of preparedness.
- Chapter 19 refers to the importance of timely post-disaster review as a source of information for the formulation of recovery programs.

## The Importance of Post-Disaster Review

4. The aforementioned references indicate the importance of post-disaster review; and there is no doubt that the experience and lessons gained from a disaster event can be used to improve many aspects of disaster management for the future. However, as explained below, post-disaster review is sometimes overlooked or otherwise omitted. The importance of ensuring, as far as possible, that post-disaster review is carried out should therefore be stressed in national policy and plans. Also, if considered necessary, such review could be made mandatory under disaster legislation or regulations.

## Reasons for Omission of Post-Disaster Review

5. In some cases, post-disaster review is overlooked or omitted from the disaster management process. Some reasons for this are given below.

- *Pressure of the recovery process*
  Sometimes recovery requirements are extensive and extremely urgent and they place heavy demands and pressures on virtually all available national resources; this can apply particularly in small countries. Post-disaster review is therefore crowded out by the need to give priority to recovery programs.

- *Government reluctance*
  If a disaster has been badly handled by government, this may invoke widespread criticism from opposition politicians and the community. A government may therefore play for time by delaying post-disaster review to a point when it is no longer useful or effective. Governments may also wish to avoid publicizing disaster management shortcomings in case these adversely affect future support from international assistance sources. These sensitivities may be understandable. However, they are counterproductive for effective disaster management and governments should recognize that a confidential unpublished report, which tells the truth to all people who need to know, is better than no report at all.

- *Administrative delay*
  Similarly, but because administrative delays or shortcomings, post-disaster review may be left until too late to be of use.

## Aspects Covered in Post-Disaster Review

6. Normally, post-disaster review needs to be as comprehensive as possible and to include the following aspects:

- Status of plans, mitigation measures, and preparedness prior to the disaster;
- Communications;
- Warning, including origin(s), transmission and receipt, processing dissemination, action taken (by government, the community, etc.), functioning of warning systems;
- Activation of the counter-disaster system and mobilization of resource organizations;
- Emergency operations centers, acquisition, receipt and handling of information, display and assessment of disaster situation, decision making, dissemination of decisions and information;
- Assignation of tasks to organizations involved in the disaster operation;
- Operations carried out, including search and rescue, casualty handling, initial relief measures, clearance of vital routes/areas, evacuation, restoration of services;
- Arrangements for emergency feeding, health, welfare, and shelter;
- International assistance arrangements;
- Assessment of public awareness programs, in the light of community reactions;
- Training aspects;
- Provision of information for recovery programs;
- Any special factors raised by the nature and effects of the particular disaster; and
- Research requirements revealed by the disaster.

7. If circumstances are appropriate, post-disaster review can include briefing from specialists on future trends and developments. This can help achieve optimum use of post-disaster experience.

## Major Investigations into Disasters

8. Following some disaster events, governments may consider it necessary to hold a major review or investigation. This can apply if liability or litigation is involved. Some relevant disaster management points are:

- In major investigations, such as legal inquiries, it may take considerable time before the outcome is released. Thus, any relevant disaster management guidelines may not be available in time for the next anticipated disaster (e.g., seasonal flooding).

- For in-country reviews which are primarily concerned with matters such as mitigation, preparedness, and response, it may be advisable for the committee concerned to issue some form of interim advice letters to government. This enables major issues to be highlighted and dealt with ahead of the findings. This is important where disasters are seasonal, especially to implement mitigation and preparedness measures before the next high-risk period.
- Where possible and permissible under the terms of reference of committees of investigation, a disaster management official should be available in an advisory role.

## Use of Results

9. The results of post-disaster review can be used in many ways. These may include:

- The amendment or revision of the national disaster plan, other plans or standard procedures. Post-disaster review can, in fact, provide a good opportunity to rewrite plans and/or procedures.
- Amendment to preparedness, mitigation, and similar measures and/or the introduction of new measures.
- Changes to organizational structure.
- Revision of or adjustment to major disaster management aspects, such as training and public awareness.
- Adjustments to national development plans.

# PART VI

# Disaster Management Support Requirements

*Chapter 21*

# Training

## Purpose

1. The purpose of this chapter is to outline the major considerations which apply to disaster management training.

2. The chapter covers:

- individual country requirements,
- the need for training,
- training policy,
- types of training and trainee categories,
- implementation of training,
- public education, and
- annex containing examples of in-country programs.

## Previous References to Training

3. Previous references to training in this handbook are summarized below.

- Chapter 4, which covers major requirements for coping with disaster, contains information on training aspects.
- Chapter 5 refers to the use of the disaster management cycle within training programs.
- Chapter 9 contains references to international assistance in training.
- Chapter 11 refers to the organizational requirements necessary to facilitate training.
- Chapter 12 recommends the inclusion within plans of a section on training requirements and refers to the use of training activities as a means of maintaining the viability of plans.
- Chapter 16 refers to the importance of training as a component of preparedness.
- Chapter 20 recommends that post-disaster review should cover lessons to be learned with regard to training.

## Individual Country Requirements

4. The training requirements of individual countries are likely to vary. For instance, these requirements will be conditioned by:

- the size and geographical disposition of the country,
- the disaster threat,
- national policy toward disaster,
- the nature and availability of disaster-related resources,
- the degree of dependence on international assistance, and
- training capability.

Countries should, therefore, carefully examine their individual circumstances and clearly establish their training requirements; also how these requirements can be met.

## The Need for Training

5. Disaster management, as it is currently developing, covers a wide range of functions and skills. These include:

- Planning;
- Organization;
- Day-to-day management activities;
- Counter-disaster operations;
- Crisis management activities;
- Logistic functions;
- Recovery management;
- Participation in major programs such as national development, prevention, and mitigation; and
- Specialist skills applicable to rescue, first aid, assessment, emergency relief and welfare, communications, information management, and so on.

6. Not all the above functions and skills call for specialist full-scale training because a basis of competence often already exists. For example, it is generally acknowledged that governments carry the major responsibility for dealing with disaster and that government resources must be utilized to optimum effect. Such resources (that is, government departments and agencies) are already trained and proficient in their basic roles. Thus, from a disaster management viewpoint, there is no need to train, say, the personnel of a public works department to clear debris from roads; they can do this as a normal procedure. Similarly, members of standard emergency services, such as police, fire

authorities, ambulance service, and others, are well trained in various crisis functions. They do not need additional training in their specialist roles. Again, government officials are practiced in day-to-day management; they do not require training in more of the same.

7. Clearly, therefore, training needs must, first of all, take into account this existing basis of competence. Only then can an accurate assessment of these needs be achieved. In principle, such an assessment will tend to indicate that certain areas of training need to be covered. These areas, which are outlined in paragraphs 10–11 below, obviously need to be compatible with training policy.

## Training Policy

8. In the past, counter-disaster training has often been regarded as a complex and expensive undertaking, which many countries could not afford. In more recent years, however, it has become clear that by using overseas training assistance and making full use of in-country capabilities, most countries can achieve reasonable training standards. In any event, three important policy considerations are usually applicable to disaster management training. These are outlined in Chapter 4 and are restated here for ease of reference.

- First, countries should identify their own training needs. To take examples from other countries and academic sources and try to apply them as exact "blueprints" are unlikely to prove satisfactory.
- Second, training programs must be designed to be compatible with and give support to counter-disaster plans.
- Third, responsibility for training must be clearly defined.

9. Chapter 4 contains a suggested outline for a simple training policy document. As an alternative, some countries prefer to state such policies in:

- A document outlining national disaster management policy, or
- A national disaster plan.

## Types of Training and Trainee Categories

10. An assessment of training needs will indicate the broad types of training which need to be undertaken and the categories of people who need to be trained. These are likely to comprise:

- *Disaster management training*
  Training for existing or potential managers (e.g., government officials and senior persons in NGOs) to equip them for specialist

disaster-related tasks, and to orient them to various aspects of crisis management.

- *Skills training*
  Training for those persons who may be required to undergo duties in:
  Emergency operations center,
  Rescue,
  First aid,
  Emergency feeding and welfare,
  Communications,
  Needs and damage assessment.

Such persons may include members of government agencies, NGOs, and community volunteers. In some cases, they may need to be trained as auxiliaries to support existing emergency services (e.g., for auxiliary rescue teams acting in support of the police and fire authorities).

- *Coordination training*
  Training for coordinated disaster management action is likely to be required for all key persons in:
  Standard emergency services,
  Government departments and agencies,
  NGOs, and
  Community/private sector groups.

  As an adjunct to this training, the services and the organizations themselves need periodic practice in coordinated action, usually in the form of combined exercises.

- *Specialized training*
  Training, often by means of workshops or seminars, to cover:
  specialized subjects such as mitigation measures,
  special briefings (e.g., politicians), and
  annual preparedness reviews.

11. While the four types of training requirements mentioned above are likely to comprise the core activities, other possibilities should not be overlooked. In practice, training requirements tend to fluctuate considerably, the emphasis on some activities fades, while the need for other activities increases. Thus training programs need careful monitoring, with adjustment being applied as necessary It should also be borne in mind that, to meet local needs, training activities could comprise a mixture of the four above-quoted categories. For instance, a course dealing with response operations might include:

- Managers,
- Persons involved in the warning process,
- EOC action desk officials,
- Persons responsible for needs and damage surveys,
- Public relations officials.

See also the appendix to this chapter which covers some possible examples of in-country training programs.

## Implementation of Training

### 12. *General Considerations*

Before establishing and implementing training programs, it is prudent to identify and consider certain in-country factors which affect training overall. These may include:

- The basis of organization which applies to disaster management, since this influences levels of training. It also provides a suitable training structure or framework.

- The current status of disaster management, because this bears on the required scope of training programs.

- The degree to which the disaster management system is utilized in some or all of its various functions (especially mitigation, preparedness, response, recovery). This will reflect existing, up-to-date experience and thus help identify types of training which need to be given particular emphasis.

- The experience of disaster management staff, especially that of key officials, which needs to be melded into and within training programs.

- The facilities which are available to implement training programs, particularly in terms of instructional staff, also accommodation, instructional aids, and administrative support (see also paragraph 16 below).

### 13. *The Aim(s) of Training Programs*

After considering factors such as those mentioned above, it is important to define the aim of training program(s) and of the courses and other activities within such programs. Clear definition of the aim helps significantly in determining the size and shape of programs and the syllabi for individual courses and activities.

## 14. *Responsibility for Training*

In paragraph 8 above, it is advocated that responsibility for training must be clearly defined within the overall training policy. In most countries, it is probably best to place this responsibility with the chair of the NDC or committee. Day-to-day training matters can then be delegated by the chair to the NDMO (or Section).

## 15. *Duration of Training Programs*

As stated earlier, training requirements tend to fluctuate. It is therefore advisable to review the contents of training programs regularly. Depending on the size and nature of such programs, this might be done annually or biennially. If a training program is being introduced for the first time, it is useful to run it over a pilot period of, say, 1–1.5 years, after which it should be evaluated as a guide for the future.

## 16. *Training Staff and Facilities*

The implementation of training programs is obviously influenced by the availability of suitable accommodation, institutional facilities/equipment, and staff. On this aspect, the following factors are usually worth considering.

- The NDMO/section should be given the responsibility for overseeing day-to-day training matters.
- Members of this office/section should receive training at a suitable disaster management establishment (e.g., the Asian Disaster Preparedness Center in Bangkok). If such training could include instructional techniques, this is advantageous.
- The NDMO/section should be provided with a set of local training guidelines. If considered appropriate, this can be done as an attachment to the training policy document, or as an annex to the national disaster plan (the latter is a useful method since the plan should come under regular scrutiny either in a post-disaster or periodic review).

- Local disaster-experienced officials and the resources of government training sections should be used to assist in formulating and implementing training programs. Also, there is usually a good deal of instructional expertise within police, defense force, educational, Red Cross, religious, welfare and similar resources; and the use of this expertise is not only invaluable from a purely training viewpoint since it also helps significantly in disaster management coordination. Astute use of these local personnel resources can go a long way to offsetting the often-perceived shortage of suitable trainers. For instance, in one country, the graduates of a series of disaster management courses for government officials formed themselves into an association of trainers.

- The use of mobile training concepts can also be beneficial. The roving seminar type of activity is not only economical in utilizing resources; it also has the additional asset of taking training direct to the users in their own area.

- Use can also be made of any systems which may be applicable to local circumstances. For instance, community radio listening groups exist in some remote areas; also, radio is widely used for disseminating government information programs (e.g., on child welfare). These radio links can similarly be used for passing training information down to the community level.

## 17. *International Training Assistance*

International assistance should never be regarded as a substitute for in-country training but it can certainly provide valuable support for the latter. The following points are therefore worthy of consideration.

- Agencies such as UNDRO, USAID, AIDAB, Red Cross, World Meteorological Organisation, World Health Organisation, and others have sponsored seminars and workshops on various aspects of disaster management.

- Certain church bodies and welfare organizations have significantly assisted their in-country branches in promoting disaster-related capability.

- AODRO (the Australian Overseas Disaster Response Organisation) has specialized in assisting NGOs, primarily in the South Pacific region. Assistance has been provided in the form of in-country workshops, aimed largely at coordinating NGO support to governments in accordance with national disaster plans.

- The Pacific Islands Development Program has included training assistance in its general disaster management support for the Pacific region.
- The Asian Disaster Preparedness Center of the Asian Institute of Technology, Bangkok, provides in-country support on various aspects requested by countries. In addition to general training, some specific technical subjects are disaster-resistant building techniques, earthquake engineering studies and cyclone warning systems.
- Countries such as the United States, Japan, United Kingdom, Australia, New Zealand, Canada, and the Netherlands have contributed in various ways to in-country training initiatives. This has sometimes been done within major international development programs.

18. ***Examples of In-Country Programs***

Some examples of possible in-country programs are given in the Annex to this chapter.

## Public Education

19. It is widely accepted that the broad objectives of training are to teach people to carry out specific tasks based on accepted methodology. Education, on the other hand, is more concerned with the development of mental capacity and thus with people's attitudes.

20. From a disaster management viewpoint, therefore, there is benefit in contributing to public education, as far as possible, in support of training programs. This contribution could also be seen as furthering public awareness, which is the subject of Chapter 22.

21. It is suggested, therefore, that national authorities concerned with disaster management training, and associated public awareness should take advantage of media and other opportunities to apprise the public of current and proposed disaster-related activities. Assistance in educating the public to understand the benefits of long-term mitigation, especially in relation to national development and preservation of the environment, should be regarded as a valid and productive objective for disaster management. In this regard, programs of disaster education in schools can obviously play an important part, as has been well illustrated in a number of countries.

# Annex to Chapter 21

## EXAMPLES OF IN-COUNTRY TRAINING PROGRAMS

1. This Annex gives some examples of possible programs which might be adapted for use as part of in-country training schemes.

2. A useful way of monitoring in-country training requirements and progress is to do so against the following categories of activity.

*Individual Training*
- Skills training (e.g., in categories such as rescue, first aid)

*Collective Training*
- Seminars and workshops (e.g., Annual Preparedness Seminar, Post-Disaster Review Workshop)
- Training modules (e.g., on General Disaster Management Response Operations)

*Collective Exercises*
- Simulation exercises (indoor)
- Skeleton exercises (outdoor)
- Unit (single agency, full scale)
- Combined

3. ***Example of a Seminar Activity: National Disaster Policy Seminar***

*Aim*
To review aspects of the national policy for dealing with disasters and to make recommendations as required.
*Method*
Briefings, presentations, discussion, and formulation of recommendations.
*Duration*
1 day
*Timing*
As required but probably annually.
*Content*
Selected aspects of mitigation; preparedness; plans; organization; response operations; training; public awareness; results of post-disaster review; other relevant aspects.
*Attendance*
Senior government and nongovernment officials holding, or likely to hold, senior positions of responsibility in disaster-related activities.

## 4. *Example of a Workshop Activity: Disaster Response Workshop*

*Aim*

To consider factors which apply to disaster response operations and to make recommendations as required.

*Method*

Briefings, presentations, discussion, and formulation of recommendations.

*Duration*

3 days.

*Timing*

As required.

*Content*

Current plans and organization; warning; activation for response operations; communications; roles of EOCs; acquisition and processing of information including survey and assessment; emergency relief activities; coordination; public cooperation.

*Attendance*

Members of government and NGOs or other persons who may have a responsibility in disaster response operations.

## 5. *Example of a Training Course Activity: Disaster Preparedness Training Course*

*Aim*

To study disaster preparedness and the implementation of preparedness measures.

*Method*

Presentations, discussions, and exercises.

*Duration*

A series of weekend or evening sessions.

*Timing*

As required.

*Content*

Current plans and organization; responsibilities for preparedness at the national, provincial, and local government levels; types of preparedness measures; warning arrangements; action on receipt of warning; public awareness and cooperation; any other preparedness and readiness factors.

*Attendance*

Members of government and NGOs or other persons who may have responsibilities relating to disaster preparedness.

*Chapter 22*

# Public Awareness

## Purpose

1. The purpose of this chapter is to outline major factors which apply to public awareness.

2. The chapter covers:

- the importance of public awareness,
- responsibility for public awareness programs,
- main aspects of programs,
- information format,
- information channels,
- organization of programs,
- effectiveness of information, and
- maintenance of awareness levels.

## Previous References to Public Awareness

3. Previous references to public awareness in this handbook are summarized below.

- Chapter 2, which deals with the disaster threat, repeatedly includes public awareness and education among general countermeasures.
- Chapter 3, on national disaster management policy, refers to the need for community involvement and awareness to develop preparedness and self-reliance.
- In Chapter 4, on major requirements for coping with disaster, the need for community involvement and participation is emphasized.
- Chapter 5 advocates use of the disaster management cycle in programs of public education and awareness.
- Chapter 8, which deals with counter-disaster resources, includes the need to cover public awareness.
- In Chapter 10, on leadership, the aware, alert, and informed community is cited as an enormously strong asset for coping with disaster.

- Chapter 11, dealing with organization, emphasizes the need to cover the development of public awareness and thus constructive community involvement and participation in preparedness, response, and recovery.
- Chapter 12 advocates the inclusion of a section within plans to cover public awareness.
- Chapter 14 quotes the need for programs of public awareness and education to keep communities apprised of risks and vulnerabilities, so that they will understand the need for preventive countermeasures.
- In Chapter 16, emphasis is placed on public awareness and education as a measure of preparedness.
- Chapter 17 stresses the need for public information during response operations.
- Chapter 19 emphasizes the importance of public information during recovery programs.
- Chapter 20 advocates the need to cover public awareness programs during post-disaster review.
- Chapter 21 suggests that public education should be broadly linked to training programs.

## The Importance of Public Awareness

4. Wide international experience illustrates and emphasizes the importance of public awareness in coping with crisis situations. Indeed, the references quoted above underline the strong connection between public awareness and virtually every other aspect of disaster management.

### 5. *The Aim of Public Awareness Programs*

The aim of public awareness programs is to promote an informed, alert, and self-reliant community, capable of playing its full part in support of and in cooperation with government, in all relevant disaster management matters.

### 6. *Important Community Needs*

In line with the above aim, most countries recognize that community members need to know the following facts in relation to possible disaster impact.

- What disaster will do;
- The best immediate action to take, personally and by families and other kin groups;

- How best to help other members of the community;
- What the government has planned to do to help the community;
- How to participate effectively in the disaster communication and warning process; and
- How to improvise shelter and sustenance until assistance is available.

### 7. *Government/Community Understanding and Cooperation*

In conjunction with the foregoing community crisis needs, there are several factors affect good understanding and cooperation between government and communities. These usually include the following:

- In addition to knowing what government has planned to do to help it in time of disaster, the community should also have at least a broad understanding of the scope and limitations of government responsibility. This applies to all disaster-related matters, including prevention, mitigation, and national development.
- The government and the community must act together to overcome the crises which arise from disaster and to ensure a rapid return to normal conditions.
- The community needs to understand and be able to implement certain measures of self-preparedness when required to do so.
- Similarly, it is important for the community to understand the government's problems in providing post-impact relief, especially the difficulties in accurately establishing immediate needs.
- It is usually in the community's own interests to cooperate, to the maximum extent possible, with the government in disaster management programs.
- For its part, the government needs to understand and take into account the fact that, under disaster conditions, the public is usually affected by shock, loss, and other trauma, and needs treating accordingly.

## Responsibility for Public Awareness Programs

8. As with training, responsibility for public awareness programs needs to be clearly defined. It is probably best in most cases to place responsibility with the authorities which also cover training. In other words, overall responsibility should rest with the chair of the NDC or Committee, with day-to-day matters being delegated to the NDMO (or Section).

## Main Aspects of Programs

9. Normally, the following main aspects of public awareness programs need to be clearly established:

- The information which needs to be communicated to the public,
- The format in which the information is to be communicated, and
- The medium (or channel) which is to be used to convey the information to the public.

10. In establishing these main program aspects, it is beneficial to consider to what extent government/community factors, such as those mentioned in paragraph 7 above, can and should be reflected in the programs themselves. Other factors which need to be taken into account are:

- Community experience of disaster, and
- Expectancy and dependency factors, which may have led communities to become overdependent on government assistance, thus eroding community and individual self-reliance.

### 11. *Program Themes*

If considered appropriate, various themes can be used in implementing public awareness programs. These, which can be changed and/or given particular emphasis from time to time, might include the following:

- Disasters are not discriminative. Certainly they affect individuals, families, and communities. However, they also affect the national infrastructure, the national economy, and the efforts of government in improving national development and the living standards of the people.
- The community and the government are interdependent in coping with disaster. They must work together to overcome the problems which arise and to restore things to normal.
- The community is better equipped to cope with disaster if it is able to undertake certain basic measures for preparedness, self-reliance, and self-help; and if it accurately understands the aims, scope, and limitations of government assistance programs.

### 12. *Information to be Communicated to the Public*

The information to be communicated to the public can be divided into various categories, for example:

*Basic Community Needs*
These can be along the lines of those stated in paragraph 6 above.

*Information on Government Assistance Programs*
This might include:

- Community-relevant details of the national disaster management system,
- Any necessary information to amplify the foregoing basic community needs. This might include special warning signals, designated shelters of safe havens, evacuation procedures, and so on.
- The limitations imposed on government assistance by factors such as the need for survey and the time this takes, logistic problems in providing relief supplies, and the necessity to establish priorities for various relief measures.
- Any other information applicable to particular circumstances.

*Seasonal Preparedness Reminders*
When probable disasters are seasonal (e.g., floods, tropical cyclones), the preseason period provides a receptive time for emphasizing public awareness factors especially, say, preparedness cleanup, warnings, possibility of evacuation, etc.

*Post-Impact Information*
Communication with the public immediately following disaster impact is a very important part of disaster response. It should therefore be a part of public awareness programs to explain:

- The general response intentions of government;
- The kind of information likely to be passed to the community;
- How it is intended to pass this information (i.e., by broadcast or other means); and
- Community responsibility for taking action on receipt of this information, when required.

## Information Format

13. The format for communicating information to the public (as distinct from the channels used) can vary. Some possible options are:

- Notices;
- Posters;

- Cartoons;
- Photographs;
- Films, film clips, videos;
- Inserts into ongoing programs on other subjects;
- Short radio or television features; and
- Talks or presentations.

## Information Channels

14. The following are examples of channels which have been used in various countries to communicate information:

- Telephone directories;
- Shopping bags;
- Existing government programs (on, say, health education), which can be used to carry short inserts concerning preparedness and other disaster-related matters;
- Radio and television;
- Cinema programs (for instance, a short film of a previous disaster is a very good reminder, especially in a preseason period).
- Special information displays (especially during a National Disaster Preparedness Day, or similar occasion);
- Police information channels;
- Use of popular sporting events, etc. to display posters or banners.
- Community gatherings/meetings of various kinds;
- Print media;
- Church, mission, and similar sources;
- Voluntary organizations, especially if they are coordinated under some form of disaster support system; and
- School programs.

## Organization of Programs

15. In organizing public awareness programs it is usually advisable to seek advice from information specialists (e.g., the government information service). As a guide, the initial organization of a program might concentrate on:

- A series of radio broadcasts (assistance with these can usually be provided by the national broadcasting authority, which has relevant experience in this field);
- Publicity in the print media (the cartoon strip method has proved effective in a number of cases);

- Use of government-sponsored community schemes (as "carriers" for disaster-related information); and
- School programs, which offer good long-term dividends, especially in developing an informed, alert, and self-reliant community.

## Effectiveness of Information

16. It is important to check, where possible, the effectiveness of public awareness programs. This particularly applies to the ability of programs to gain and maintain the interest of the people who are being targeted. It is also advisable to check periodically that the information being imparted by programs is being received by the public in its intended sense.

## Maintenance of Awareness Levels

17. The maintenance of adequate awareness levels needs to be continuously monitored. Otherwise, programs will likely become stale and public interest will fade.

18. In Chapter 12 on plans, a definite program, or series of programs, is recommended to be established to help in maintaining the viability of plans. Public awareness activities are also suggested to be linked into such a program. This kind of approach probably provides the most effective basis for maintaining awareness levels. However, it is also suggested that a watch be maintained for any innovative ideas which may help to renew or freshen up awareness interest. Some of these ideas, while basically simple, can be very effective. For instance, manufacturers and distributors of everyday household goods can often be persuaded (in their own as well as the community interest) to use some promotional gimmick which will have strong public appeal. Overall, however, watchfulness by the responsible disaster management authority is the best insurance for maintaining public interest.

*Chapter 23*

# Research

## Purpose

1. The purpose of this chapter is to provide a short reference to aspects of research which are of particular relevance to practical disaster management.

2. The chapter covers:

- The extent of modern disaster-related research.
- The importance of research.
- Practical use of research information.

## The Extent of Modern Disaster-Related Research

3. If one considers the scope of all disaster-related activities, it is clear that this handbook represents only a fraction of the total field. Yet the foregoing chapters of the handbook cover many differing aspects—from hazards to organization, from planning to leadership, from mitigation to human behavior. It is not surprising, therefore, that disaster-related research can extend across a literally enormous area.

4. Indeed, if one looks at the contents list of this book, fields of matching research are readily apparent. Some existing examples are:

*Meteorology and Climatology*
  Climatic change, various aspects
  Flood forecasting
*Earthquake*
  Seismicity and earthquake risk
*Volcanic Eruption*
  Volcanic hazards, various aspects
*Cyclone*
  Wind tunnel tests
  Fatigue loading on houses

*Drought*
Rainfall patterns
Desertification
*Chemical Hazard*
Air pollution
Chemical hazard
*Organization and Management*
Organization response
Behavioral aspects
Flood hazard management
*Planning*
Land-use planning
Hazard mapping
*Mitigation*
Flood mitigation strategies
Shelters
*Public Awareness*
Public participation in counter-disaster activities

5. In addition to what might be called specific disaster research, such as the areas quoted above, it is important not to overlook lesser, but nonetheless valuable contributions. These might include investigation into:

- Specialized aspects of transport safety (e.g., air traffic control facilities which may help guard against a major aircraft accident);
- The application of legislation to aspects which can contribute to disaster mitigation;
- Investigation into and development of arrangements, such as insurance incentives, which help governments, businesses, and individuals reduce effects and losses.

6. While the extent of current research is encouraging, instances are repeatedly quoted of research efforts being constrained by limitations of funding, low priority being given to disaster-related research projects, and disaster-related research having to be "back-carried" by larger research projects.

These constraints underline the importance of the relatively few centers which undertake specialized disaster research and those disaster management institutions which have a research responsibility within their charters.

7. An additional factor is the current limitation which applies to post-disaster review and the evaluation of relief (and other) programs. Post-disaster review is covered in Chapter 20. However, with regard to the evaluation of

disaster-aid programs, Cuny (in Disasters and Development) notes that there is a marked reluctance by the agencies concerned to do this. He maintains that there are a number of reasons for this, including the fact that if an evaluation is negative, and this information is made public, future agency funding may be affected. As Cuny notes, failure to evaluate programs means that mistakes are repeated again and again; thus the basis of information from which improvements might flow remains undeveloped.

8. From a disaster management viewpoint, therefore, the overall situation concerning research might be summarized as follows:

- There is a wide range of research into disaster-related subjects but projects are often constrained by restrictions on priorities and funding.
- Specialized disaster research is undertaken by only a very limited number of institutions.
- Post-disaster review and program evaluation make some contribution to research efforts, through the process of lessons learned, but this contribution needs to be strengthened (e.g., by identifying topics which require research).

## The Importance of Research

9. The importance of research in relation to disaster management is that it should:

- Help eliminate the repetition of previous mistakes,
- Contribute to improvements in ongoing disaster management capability,
- Assist in reducing vulnerability to disasters, and
- Stimulate forward-looking concepts for the future.

Toward these ends, the results of research must be usable and used. If they are not usable, the major onus is on the researcher. If they are usable but not used, then the disaster manager is at fault. There will obviously be peripheral exceptions to these two assertions but, fundamentally, they lie at the heart of the relationship between the researcher and the disaster manager. There is, of course, the added factor (and it is a critical one) that there must be coherent dialogue between the two. To establish this effective dialogue is something to which both the researcher and the disaster manager might give greater attention in the future. For certainly in the past, it has all too often been evident that the gap between the two was counterproductive for all concerned.

10. However, it needs to be emphasized that the information flow is very much a two-way affair. The disaster manager needs the researcher's information if he/she is to improve things for the future. At the same time, the researcher needs feedback from the disaster manager to ascertain whether his/ her research information has worked; what effect developing disaster management practice is likely to have on upcoming research needs; and what kind of information, therefore, is required by the disaster manager in the future.

## Practical Use of Research Information

11. In view of the foregoing factors, it is suggested that disaster managers should take a planned approach toward use of research information, rather than wait for occasional user opportunities to occur. Major factors in such an approach are likely to be:

- *Disaster experience*
  This, especially if it includes post-disaster review, should indicate broad disaster management areas in which adjustments and improvements are necessary. Such adjustments and improvements, if they are to be viable, need to be based on the best research and other information.

- *Disaster management review aspects*
  In addition to actual disaster experience, other aspects of disaster management act as a form of review. These include national reports (e.g., annual report to government by a National Disaster Council), training programs (particularly exercises and tests), and periodic checks of plans and operating procedures. The results of such reviews need to be applied in the light of current research knowledge and information.

- *Projections of major national programs*
  National 5-year plans (or similar major planning initiatives) require the best available existing data and forecast projections. Many countries now use some form of integrated approach to policies covering national development, environment protection, and disaster management. Up-to-date disaster research information can obviously make a valuable contribution to the development of these national programs.

- *Major developments within national disaster management*
  Initiatives which prompt developments within national disaster management may arise from various sources (e.g., disaster events,

government decisions, public pressure, standard reviews). Sometimes these initiatives may constitute major changes from past policy; for instance, a country which has previously concentrated mainly on preparedness, response, and recovery may deem it necessary to embark on a major program of mitigation. Such a program may involve large sums of national and international assistance money and may depend critically on specialist, technical, and other information. In such cases, there is an obvious need to obtain the best possible up-to-date information as a program base. This need may have to be met by some form of specific research input; for example, a specialized review of the proposed program against a comprehensive background of available research information.

- *Effects of specialist research efforts*
  Sometimes a breakthrough in specialist research effort may have a significant effect on disaster management arrangements. For instance, advances in communications systems may be so important in terms of warning capability and public reaction that they may have to be given near-immediate priority over other systems or facilities.

- *Monitoring of available information*
  Available research information obviously needs to be monitored by disaster management authorities to ensure that relevant data can be assessed and, where appropriate, used. Disaster managers therefore need to access various available mailing lists, such as the *Natural Hazards Observer* (published by the Natural Hazards Research and Application Information Center, University of Colorado, USA) and other publications. In this regard, it is of note that during 1990, the Natural Hazards Research and Applications Information Center advertised a Quick Response Program to enable disaster specialists to conduct short-term field investigations immediately after impact (thus avoiding loss of valuable initial data). This monitoring role should be written into the duties statements of appropriate officials/ organizations (e.g., a national disaster management office).

- *Indirect sources of research information*
  International conferences of various kinds often provide valuable access to research information and its use. Attendance at such conferences by disaster management officials should therefore be encouraged, especially because it may throw valuable light on the success or otherwise of efforts to use research.

12. The foregoing examples represent only some of the approaches which might be made toward use of research information. Obviously, therefore, disaster management authorities need to develop their own system.

# Appendixes

*Appendix A*

# Vulnerability Analysis

## Purpose

1. The purpose of this appendix is to outline the major factors which apply to vulnerability analysis.

2. The contents of the appendix are particularly relevant to:
   - Chapter 2 which deals with the disaster threat, and
   - Chapter 15 which covers mitigation.

## Acknowledgments

3. The entire contents of this appendix have been taken from the *Technical Background Paper of Disaster Mitigation in Asia and the Pacific* (1990), prepared by the Asian Disaster Preparedness Center of the Asian Institute of Technology, for the Asian Development Bank. The authors of this paper are Dr. Ian Davis and Professor Dr. S. P Gupta. This contribution to the handbook is gratefully acknowledged.

## Contents of the Appendix

4. The appendix includes the following main aspects:

- Risk,
- Hazard assessment,
- Vulnerability analysis,
- Physical vulnerability,
- Social vulnerability,
- Economic vulnerability,
- Resource requirement,
- Decision making, and
- Defining an acceptable level of risk.

## *Risk*

5. Risk is defined as the probability of meeting danger or suffering harm or loss. In relation to disaster, risk has been more specifically described as the probability that a disaster will occur, using relative terms such as high risk, average risk, and low-risk to indicate the degree of probability. Risk assessment includes an evaluation of all the elements that are relevant to an understanding of existing hazards and their effect on a specific environment.

6. Knowledge of hazards is provided by the various physical sciences such as meteorology, hydrology, geomorphology, seismology, and volcanology. The understanding of vulnerability includes physical, social, and economic aspects. To turn risk assessment into a useful planning tool, accurate information must be gathered from numerous and diverse sources, ranging from remote sensing of crop yields to the monitoring of volcanic activity, to the historical records of floods or earthquakes, and social surveys of livelihoods, living patterns, and a community's perception of risk.

7. From these data, a fairly complete picture may be assembled of the prevailing risk expressed in terms of probabilities. There are several steps in risk assessment based on the related processes of hazard mapping and vulnerability analysis. They establish the nature, location, and scale of risks to society and its assets. This information can assist decision makers in deciding what can and should be protected and up to what level of safety, measured against the potentially disastrous effects of not initiating risk-reduction measures.

## *Hazard Assessment*

8. Disasters result from vulnerable conditions being exposed to a potential hazard. Therefore, the first step in taking any mitigation measures is to assess the hazard. Hazard assessment aims to come to grips with:

- the nature, severity, and frequency of the hazard;
- the area likely to be affected; and
- the time and duration of impact.

Hazard assessment concerns the properties of the hazard itself (i.e., cyclone, flood, drought, earthquake, volcanic eruption, etc.) and its direct effect, not its effect on the socioeconomic environment which is covered by vulnerability analysis.

9. Hazard assessment begins with data collection; existing assessments and hazard maps; scientific data (meteorological, hydrological, seismological,

volcanological, etc.); other maps (topographical, geological, etc.); local lore and historical records; and socioeconomic or agricultural surveys. The data are then analyzed. One effective way of presenting hazard assessment is through hazard mapping. Hazard maps can be of macro or micro scale; for cyclones (cyclone track maps, wind velocity zoning maps, and cyclone/storm surge hazard maps); for floods (inundation maps); for earthquake and tsunami (epicentral maps, seismic zoning maps, seismic zone source maps, tsunami hazard maps); and for landslides (landslide hazard maps).

10. Next, data and maps are analyzed to arrive at the hazard assessment. The level of sophistication of hazard assessment will depend on the perception of risk and available resources. Formats for expressing the intensity (power or impact) of a particular hazard are well established. Tropical cyclones can be measured in terms of wind speeds; floods in terms of return periods, duration, and inundation levels relative to normal river or sea levels; earthquakes in terms of intensity indicating their effects on man-made structures or in terms of magnitude on the Richter scale.

11. Finally, an attempt should be made to express the probability of the event occurring over time. Probabilities are assessed on the basis of scientific data and historical records. Sophisticated new tools are available to assist in hazard mapping and assessment. For instance, aerial photography and satellite imagery can provide information about the landscape, flooding, fissures, fault lines, etc. Microcomputers have a growing potential for applications in developing countries as they become more powerful and relatively inexpensive. They can be used to store and present geographical data employing Geographic Information System (GIS) techniques in preparing hazard maps and for hazard modelling. In the simplest possible terms, hazard assessment attempts to specify that there is a certain probability of an event of a given proportion occurring in a certain area. It cannot predict with certainty when this event will happen or what its effects may be.

## *Vulnerability Analysis*

12. Having established the space/time/intensity dimensions of hazard incidence as well as its general characteristics, the second step is vulnerability analysis. This is the process used to identify vulnerable conditions which are exposed to natural hazards. If an area is exposed to multiple hazards, vulnerability analysis should be carried out for each type of hazard. Like hazards themselves, vulnerabilities can be conveniently depicted on maps, either as a single vulnerability map for each type of hazard or as a multi-hazard vulnerability overlay. In general, vulnerability analysis provides information on:

- the sectors at risk: physical (buildings, infrastructure, critical facilities, agriculture); social (vulnerable groups, livelihoods, perception of risk, local institutions, poverty); economic (direct losses, indirect losses); and
- the type of risk (damage to public infrastructure, production facilities, housing, or casualties).

Vulnerability analysis results in an understanding of the level of exposure of persons and property to the various natural hazards identified.

## *Physical Vulnerability*

13. Physical vulnerability relates to buildings, infrastructure, and agriculture. The major focus of this appendix is on physical assets but the loss potential of crops, trees, livestock, and fisheries should be recognized as well.

The vulnerability of buildings is affected by their site, design, shape, materials used, construction techniques, maintenance, and proximity of buildings to others. The weighing attached to each factor will vary according to the type of hazard encountered. Different hazards produce different forces affecting these structures.

The vulnerability of infrastructure is also specific, depending on the type of hazard. Infrastructure may be considered in three broad groups; transport systems (roads, railways, bridges, airports, port facilities); utilities (water, sewerage, and electricity); and telecommunications. Hazard protection measures such as flood embankments are also considered part of the infrastructure once they have been installed.

Vulnerability analysis is especially concerned with the risk faced by critical facilities (sometimes termed "lifeline") which are vital to the functioning of societies in disaster situations. Such facilities include hospitals and dispensaries; emergency services; key transport and communications systems; essential services and utilities; high occupancy buildings; and structures such as dams that are essential to the economy. Special consideration is given also to assessing the vulnerability of buildings of great cultural importance.

## *Social Vulnerability*

14. While the physical elements listed above have received primary attention in past vulnerability studies, social vulnerability has been somewhat neglected. A general methodology to measure social vulnerability is still being developed, focusing on special categories of vulnerable groups; livelihoods

at risk; perception of risk; existence of local institutions; and level of poverty. Records of past disasters suggest that the following groups of people are particularly at risk and require special attention:

- single parent families;
- women, particularly when pregnant or lactating;
- mentally and physically handicapped people;
- children; and
- the elderly.

People living or working in remote areas and seasonal migrant labor groups may also be at risk and pose special problems for both relief and mitigation. If people are aware of potential hazards, their nature, and their likely impact, and understand what actions need to be taken to reduce risks, then they become less vulnerable. Risk perceptions need to be assessed and awareness programs need to be initiated. There are likely to be three determining factors. First, poverty levels are important because the poorer people are, the more they are exposed to everyday risks and the less concerned they are with infrequent hazards. Second, risk perception relates directly to the level of public awareness programs. Finally, the frequency of repeated hazards determines the perception of risk. In sum, if there are groups whose livelihoods are at risk, living or working in densely populated areas, with low perceptions of risk, and without institutional support, the cumulative effect would be high social vulnerability.

## *Economic Vulnerability*

15. Economic vulnerability measures the risk of hazards causing losses to economic assets and processes. It focuses on evaluating the direct loss potential (i.e., damage or destruction of physical and social infrastructure and its repair or replacement cost, as well as crop damage and losses to the means of production); indirect loss potential (i.e., the impact on lost production, employment, vital services, and income-earning activities); and secondary effects (epidemics, inflation, income disparities, and isolation of outlying areas).[1] With the insights provided by economic vulnerability analysis, it is possible to estimate direct and indirect losses and to design ways and means to mitigate them in relation to the estimated costs of relief/recovery actions and mitigation measures required. The estimation of potential losses can be carried out using various disaster scenarios based on mild, average, or worst case assessment to allow the authorities to gauge the scale of the problem.

---

1 UNDRO. 1979. *Disaster Prevention and Mitigation. A Compendium of Current Knowledge.* Vol. 7 "Economic Aspects," p. 3, UN.

It is, however, a hypothetical exercise and therefore unavoidably imprecise. With computer-modeling techniques, greater levels of accuracy will likely be attained in the future. However, disaster modelling is currently still in its early development stage. While computer modeling techniques for damage prediction have been extensively used in industrialized countries, they are only now being linked to the social/medical sphere to produce casualty predictions, which can then be incorporated into health disaster contingency planning.

In all phases of loss assessment, the primary source of data will be information on damage and casualties suffered in earlier disasters. This highlights a need for maintaining precise records. From these data, reasonable predictions of the likelihood of future losses in a similar disaster environment can be made. The projections need to be linked to the studies of physical and social vulnerability. The data from this exercise have two kinds of user: the emergency planning staff who can use them to refine their planning, and the staff involved in general risk reduction activities who need the data to justify the expenses of mitigation programs aimed at reducing the incidence of future damage.

Information from hazard maps and physical vulnerability analysis may be used in economic vulnerability analysis. The aim is to ascertain what is likely to be destroyed or damaged. The resulting damage estimates can be crude general statements or more precise calculations. For example, in the case of buildings, the estimates could be given in three categories: slightly damaged but repairable at a cost not exceeding 20% of building value; severely damaged, but repairable at a cost 20–60% of building value; and destroyed or irreparably damaged if repair costs would exceed 60% of value. A key problem of damage estimation is the difficulty of distinguishing between normal deficiencies or weaknesses (i.e., existing pre-disaster damage resulting from poor maintenance) and damage induced by the disaster event. In practice, it is almost impossible to separate the two conditions.

In the case of commerce and industry, direct economic losses relate to buildings, plants, and stocks of raw materials and products. Indirect losses are more difficult to assess. For example, the full economic losses resulting from communications disrupted by a damaged railway track or a collapsed bridge are difficult to quantify. Industrial production figures for a given factory or agriculture region for earlier years may help identify the sudden interruption in the flow of goods which could result from a disaster event. Indirect economic losses may also arise over the longer term if markets are lost permanently because of interrupted production.

If a country receives a grant from external sources for rehabilitation or reconstruction, part of the economic losses may be offset. However, UNDRO estimates that the international aid received after a disaster seldom exceeds 4% of total losses.[2] Inevitably, therefore, the destruction caused by natural disasters means a net loss of economic resources. Their replacement requires a reallocation of resources which might otherwise have been used to further the development process. It is in terms of their opportunity cost incurred that disaster impact should be measured and that the benefits of mitigation programs should be assessed.

## *Resource Requirement*

16. After mapping hazards and analyzing vulnerability, the resources required for mitigating the risks need to be evaluated. If the risk of potential losses is great, a correspondingly high level of resources will probably be required for mitigation. The aim is to achieve a balance between risks and resources. Disaster preparedness seeks to identify these resources prior to the onset of emergency situations.

## *Decision Making*

17. Having reached the stage in the assessment process where information on hazards and vulnerability has been collected and analyzed in terms of potential losses and resources available to address identified risks, the next step is to make decisions on who or what to protect and to what level of safety. One option might be not to proceed with mitigation on the grounds that it is excessively expensive to provide protection, or that the risks are too slight to be considered. In most cases, the decision to proceed will require further analysis of the degree of protection to be provided and of the relative priorities of assets to be protected.

The procedures for decision making on mitigation matters vary between industrialized and developing countries, depending on the amount, detail and quality of background material available and the resources that may be mobilized for introducing and sustaining mitigation programs. The usual method used in industrialized countries to determine how much a country, region, or town should invest in flood protection measures is based on cost-benefit analysis. The technique is not as advanced in the case of earthquake hazard due to the large number of variables in the economic risk assessment. However, the cost of upgrading a structure to make it earthquake resistant for a particular severity of seismic event can be easily calculated where the

---

2 Zupka, D. 1988 Economic Impact of Disaster. *UNDRO News*, January/February. p.19.

necessary expertise is available, thus enabling a decision on such investments to be reached. However, very few developing countries have used cost-benefit analysis to establish standards of protection. The reasons for this include the difficulty of obtaining accurate data on direct and indirect loss potential, or the lack of economic appraisal skills. In the absence of such techniques, decisions are currently made on the basis of considered judgement and the best advice available based on practical knowledge and experience.

## *Decision Making – Economic Appraisal*

18. There are three stages in cost-benefit analysis in the context of disaster mitigation activities,[3] beginning with the enumeration of all the benefits and costs of an anticipated activity to reduce risks. Next, these costs and benefits are expressed in money terms. Finally, all future benefits and costs are discounted into present value terms. Based on this analysis, which includes a probabilistic estimate of the likelihood of the hazard, decision makers can select the option in which the present value of a mitigation proposal is positive and greater than that of available alternative actions. There are three methodological problems with these three steps. First, there are problems of quantifying many of the social, psychological, or political risks in economic analysis and assigning prices to human life, emotional stress, or political upheaval. Economists using cost-benefit analysis would find it impossible to determine the monetary value of avoiding such losses or disturbances or to measure the value of security which comes, for example, from living in an earthquake-resistant house or on flood-protected land. It is even more difficult to evaluate the political benefit derived by a government which imposes safe building codes or, conversely, the potential political cost to a government which does not.

The second problem of cost-benefit analysis is the difficulty in placing a monetary value on the loss of nonrenewable national resources such as agriculture land lost through erosion, landslides; or floods. Finally, there are methodological problems in discounting future benefits and costs, and in incorporating into the analysis the risk of natural disasters occurring. This relates to determining how to handle future uncertain outcomes in relation to a host of other factors that may affect human existence in a disaster-prone environment. There are no ready solutions to resolve the limitations on assessing the costs and benefits of mitigation measures. The lack of expertise experienced in many developing countries and the conflicting demands on scarce financial, human,

---

[3] Anderson, Mary B. 1990. Analysing *the Costs and Benefits of Natural Disaster Responses in the Context of Development*, Environmental Working Paper No. 29, The World Bank.

and institutional resources further complicate the issue. Technical assistance may be able to provide support to fill identified gaps so that political leaders grappling with the difficult question of what to protect and to what level of safety may have more precise information on options open to them. Once mitigation measures have been identified, resources would then have to be mobilized at home and abroad for implementing them.

## *Decision Making – Levels of Safety and Priorities for Protection*

19. Whether officials use more or less rigorous cost-benefit analysis or considered judgment, they will have to make decisions on determining levels of protection as well as on setting priorities. Factors that will influence their decision are:

- the value of the area, economic asset, or building being protected relative to other disaster or development priorities;
- available resources;
- nationally or arbitrarily agreed standards (in some countries flood protection measures only apply to 100-year return period events); and
- the specific aims of mitigation; for example, in the case of earthquake, a generally accepted objective in developing seismic (anti-earthquake) measures for low-income dwellings is to eliminate risk of injury and death, but accept a certain level of structural damage (this is sometimes termed "safe collapse").

Whatever the approach, whether economic appraisal or common sense, the scale of expenditure on protection measures has to be established in a given set of conditions. Thus, a comparison can be made between the cost of mitigation measures and the expected benefits stemming from them. The use of such techniques can provide guidance on what level of protection to provide. Economic appraisal may also offer decision makers some comparative information on the potential economic return of mitigation measures in different sectors, or of different mitigation measures in the same sector.

## *Defining an Acceptable Level of Risk*

20. Thus far, a wide range of disciplines will have been involved in mapping assessing, analyzing and estimating risk, hazards, and vulnerabilities. The product of these activities will be a comprehensive assessment of the causes and effects of potentially disastrous events and the provision of advice on a given course of action or a range of options for taking mitigation measures

within a set of identified constraints. The ultimate decision on the level of protection and on what or whom to protect will be political, taken in the light of available resources and the perceived importance of hazard protection relative to other pressing demands on the national or local budget.

All individuals and societies have to live with risk of various types and severity. It is utopian to imagine their entire removal. Instead, mitigation measures have to be accommodated considering a variety of trade-offs. As a final step in the risk assessment process, therefore, governments in hazard-prone environments have to make decisions on what risk-taking is deemed acceptable and tolerable on the national scale. The cost of repeatedly devoting public funds for relief and rehabilitation activities following disasters must be balanced against the cost of providing preparedness and mitigation measures before the event. The acceptability or tolerance of risk is a dynamic sociopolitical process that relates to another dynamic context, that of ever-changing hazards and vulnerabilities. What one society will tolerate will not be acceptable to another. And what is acceptable at one stage in a society's economic development or value system may change significantly by a later stage.

Four causal factors influence change in the acceptability of risk by a given society:

- The development of general education and public awareness programs producing a heightened knowledge of vulnerability and of ways to secure protection at various levels of aggregation;
- General economic progress, with people no longer prepared to see their growing assets endangered by events that their parents may have tolerated: as a result, greater demand is placed on governments for a safer environment;
- Awareness of damage potential, with governments recognizing that potential losses from disasters may pose unacceptable risks for economic welfare and their own political survival; and
- Changing patterns of hazards and vulnerabilities, with communities becoming aware that new risks are posed as rivers change their courses, volcanoes become active after years of inactivity, poor families with no alternative place to live move onto disaster-prone floodplains, and with environmental degradation threatening the very Earth.

*Appendix B*

# Warning and Warning Systems

## Purpose of the Appendix

1. The purpose of this appendix is to outline the major factors which apply to disaster warning and warning systems.

2. The contents of the appendix are particularly related to Chapters 16 and 17, which deal respectively with preparedness and response.

Specific references are:

- Chapter 16 which covers various aspects of warning and outlines precautionary measures that can be taken prior to disaster impact, if there is sufficient warning lead time.
- Chapter 17 which summarizes the main requirements applicable to warning.

3. The appendix necessarily overlaps some of the information contained in the above references. However, it aims to amplify this information and provides a ready reference to the subject of warning generally.

4. The appendix covers:

- the significance of warning,
- the disaster warning and information sequence,
- capability requirements for effective warning,
- transmission of warning information, and
- factors affecting the utilisation of warning.

## The Significance of Warning

5. The significance of warning is well summarized in the Manual of Disaster Prevention and Mitigation, Preparedness Aspects, Volume 11, issued by the Office of the United Nations Disaster Relief Co-ordinator (UNDRO), which states:

It is axiomatic that if timely warning can be given of an impending or probable event which may bring disastrous consequences in its train, then it will be possible to reduce the severity of those consequences. The degree to which this reduction can be effected will depend upon the interplay of three main elements, namely:

- The accuracy of the warning;
- The length of time between the warning's being issued and the expected onset of the event; and
- The state of pre-disaster planning and readiness.

Included within this last (element) is a sub-element, that is the degree to which the public respond to the warning and take correct precautionary action.

## The Disaster Warning and Information Sequence

6. In considering and assessing various aspects and components of warning and warning systems, it is useful to relate them to the sequence which usually applies to disaster information. A typical pattern is given below and while this might not apply to every single case, it illustrates in summarized form what is involved in providing warning of a disaster. It is interesting that this type of sequence applies to virtually all emergencies, ranging from, say, a traffic accident to a major disaster.

**Origin**

*The Crisis or Disaster Scene*
Information may be and sometimes is dormant or inactive at this stage, unless action is taken to make it otherwise; for example, amateur radio operators have sometimes been the first activators of information concerning a disaster event (other agencies having been put temporarily out of action).

**Warning/Alerting**

May be from

- Ordinary source
  (Initiated by a person/citizen involved in the crisis or on the crisis scene); or

- Special source
  By warning system; or by police/emergency service/other officials who may have a special role or responsibility; or by satellite imagery or overflying aircraft.

This stage activates and starts to channel the information.

**Transmission**

This is the process of moving the information from the crisis scene to the point where it can be assessed and acted upon.

The Methods may be
- word of mouth,
- telephone/telex/facsimile,
- radio,
- alarm signals,
- etc.

**Reception and Recording**

by some form of information center which may be
- Emergency operations center
- Police HQ
- Any other relevant point or center

The information center must normally have the capability for processing the information effectively, and especially for recording it accurately and clearly.

**Display**

This may be necessary to ensure/enhance the clear recording which may be required at subsequent stages. It is not always mandatory; for instance, in minor incidents recording alone will suffice.

Display may be done by
- Maps/charts,
- Display boards,
- Visual projection, and others.

**Assessment**

The essential "information-using" phase when information is examined with regard to its own significance and its relevance to other information or factors on hand.

May be done

- By one individual acting alone.
- By one individual acting on advice of his staff.
- In conference.

**Decision Making**

The critical phase or peak via which the information is translated into action.

The decision(s) may be made

- By the individual who carries overall responsibility for the action
- By the responsible individual in consultation with his immediate staff/specialist advisers.

**Action**

The extension of the decision making into a line of action.

It may be

- *Dynamic*, such as setting off a course of operations (e.g., a survey, a search, an evacuation, a deployment of resources, a warning or instruction to the public).
- *Static*, such as remaining on stand-by pending further information, or judging that no additional action is warranted.

Finality

The stage when all action in relation to the information is complete.

This may mean

- All clear after a warning,
- Stand down,
- End of an operational phase,
- Absorption of the particular information into a subsequent phase/edition of information.

7. The foregoing pattern or sequence is a useful indicator to a number of key requirements, if warning is to be effectively obtained and utilized. These requirements are outlined below.

## Capability Requirements for Effective Warning

8. As mentioned briefly in Chapter 16, the effectiveness of a warning system depends on certain interrelated capabilities. These are:

- *Capability to receive international warning*
  For example, cyclone warnings from tropical cyclone warning centers in various locations; tsunami warnings from the Tsunami Warning Center in Hawaii; meteorological indications from weather satellites of possibly developing threats.
- *Capability to initiate in-country warnings*
  Necessary in cases such as floods, landslides, volcanic eruptions, earthquake (if prior indicators are available), major releases of hazardous substance.
- *Capability to transmit warning from national level and other key government levels*
  Mostly done by radio links or broadcast systems.
- *Capability to transmit warning at local community level*
  May be done by local radio stations, sirens, loud hailers, bells, messengers.
- *Capability to receive warning and act upon it*
  This requires:
  - possession of or access to a radio receiver or similar facility,
  - being in hearing/seeing distance of signals,
  - knowing what various warnings mean, and
  - knowing what action to take.

9. If there are limitations or disruptions to any of these capabilities, warning may not be totally effective. Most limitations or disruptions tend to be caused by the disaster events themselves (e.g., the inability of a radio broadcast system to function because of cyclone damage or damage to power supplies, including stand-by power facilities).

10. In situations where there is no advance warning of disaster (e.g., a sudden earthquake), the warning system is automatically bypassed and disaster response has to be initiated at the post-impact stage. However, the system (or at least parts of it) usually remains available for passing vital information concerning post-impact action, including community advice and direction.

## Transmission of Warning Information

11. While some warning information may be conveyed by other means (e.g., alarm systems and visual signals), most information is transmitted by telecommunications links. For instance, teleprinter links are widely used for conveying meteorological information. Some countries possess special communications networks for disaster purposes but, even so, other networks also tend to be involved. Many countries therefore make arrangements to utilize existing communications facilities to the optimum extent for disaster purposes (including warning). A typical example would be:

- *Warning from external sources*
  Radio, telex, telephone, and in some cases satellite links, channelled to a central in-country authority; e.g., a national meteorological center or national observatory.

- *In-country transmission of warning information*
  Warning may be originated by various agencies; e.g., a meteorological center, a volcanological monitoring system, an industrial safety system, a local or regional government headquarters. Information may be passed via:
  - a discrete disaster network;
  - a police network, used for disaster operations coordination (because of its wide national coverage and, probably, its 24-hour manning);
  - a national government network;
  - NGO networks, if so arranged under official disaster plans.

- *Public broadcast/dissemination of warning information*
  The authority for and system of dissemination of warning information must be clearly laid down in disaster plans. For example, a national or regional meteorological office would normally monitor and clear the format of a cyclone/windstorm warning. Broadcast of warning would be undertaken by national/regional broadcasting systems. Other means of dissemination (e.g., by sirens, loud hailers, warning pennants) would be undertaken by police, emergency services, local government authorities, etc. in accordance with disaster plans.

- *Notification of warning to neighboring countries and international assistance agencies.*
  This may be done by various agencies/channels; e.g., diplomatic missions, United Nations agencies (especially UNDP/UNDRO),

Red Cross, international church/welfare bodies, the media. Such notification can often provide useful advance warning to outside agencies that their assistance may be required.

- *Good-neighbor assistance in dissemination of warning*
  Cases are on record where countries have assisted a disaster-stricken neighbor by broadcasting warning and other information. This is particularly valuable when the broadcast/communications capability of the stricken country has been damaged and/or disrupted by the ongoing disaster. Preparedness measures can help ensure the efficacy of this kind of arrangement.

- *Back-ups ystems*
  Earmarking and inclusion in plans and operating procedures of other networks which could be called upon for back-up purposes; e.g., networks of individual government departments, private sector networks, aeronautical and maritime networks.

## Factors Affecting the Utilization of Warning

12. In responding to disaster, much will obviously depend on the action taken by disaster management authorities and the general public following receipt of warning. Such action may be influenced by a number of specific factors related to individual circumstances and particular types of disaster. However, certain factors will tend to apply generally. These factors (some of which relate to the UNDRO quotation in paragraph 5) are outlined below.

### 13. *Warning Lead Time and Accuracy*

Good warning lead time is clearly a valuable factor in response from both disaster management and public reaction viewpoints. Under best-case scenarios, it facilitates optimum:

- Readiness;
- Pre-impact precautionary measures (see Chapter 16);
- Activation (of organization, systems and facilities, etc.); and
- Mobilization.

However, long-warning lead time may mean that original forecasts concerning the threat (e.g., its severity and timing) proved to be inaccurate, resulting perhaps in unnecessary reaction. Often, therefore, it is a matter of fine judgment by senior disaster management authorities in deciding the timing of their initial response action (see also paragraph 16 concerning warning overkill). The following issues are critical to get the best use out of available warning:

- The authorities responsible for originating warning must clearly understand the difficulties and repercussions which inaccurate warnings impose on both disaster management and the public, and should fine-tune their systems and procedures accordingly.
- Disaster management authorities should attune themselves and their operational concepts to minimal warning situations. In this way, they can likely maximize their utilization of any warning period, long or short.
- Where disasters tend to be regular (e.g., seasonal flooding), past records will provide useful warning guidelines for both disaster management and the public. In fact, the public may develop more faith in such guidelines, rather then follow ongoing forecasts; this point needs to be watched by disaster managers, especially at the local level.

## 14. *Framework of Planning and Organization*

From the preceding information in this appendix, it will be readily appreciated that many agencies and individuals are involved in the provision and utilization of warning. It follows that effective coordination is vital. This, in turn, indicates that the framework of planning and organisation within which the warning system has to operate must be kept at peak effectiveness. In fact, a good way to test this framework is to check it, step by step, against the criteria of paragraphs 6 and 8 above. This check is also useful when a disaster management organization is being initially set up.

Obviously, therefore, the main needs in planning and organisation are that:

- Plans must contain adequate information and direction concerning the coordinated utilization of the warning system; and

- The organizational structure must facilitate the best possible functioning of the warning system and cater for the provision and maintenance of the necessary facilities and systems, including communications links.

Inadequacies in planning and organization (and associated preparedness) will affect not only the utilization of warning but, of equal importance, its acquisition as well. In this regard, Chapter 16 lists some important points concerning:

- Warning systems;
- Readiness;
- Activation;
- Provision for non-warning situations;
- Testing and practice of warning systems; and
- Precautionary measures prior to disaster impact, if adequate warning is available.

## 15. *Public Awareness and Participation*

Public awareness and understanding of the warning system plus community capability to participate in related activities (e.g., precautionary measures, evacuation procedures) constitute a key link in the total warning chain. Chapter 22, which deals with various facets of public awareness, emphasizes the need for measures which will help maintain awareness levels.

## 16. *Warning Overkill*

What might be called warning overkill can adversely affect reaction to and utilization of warning. For instance, neither disaster management authorities nor the general public can be kept at a high state of standby/readiness over long periods. The use of warning must therefore be selective in the sense that it avoids, as far as possible:

- Continuous awareness warnings which eventually become disregarded (this needs to be particularly watched if disasters are seasonal, when it can be possible to overemphasize such warnings);
- Repeated false alarms because warnings are issued for very low-level threats, or because the threats are not sufficiently and accurately diagnosed before warnings are issued.

In other words, there is usually a peak in warning/readiness/public awareness beyond which effective reaction will rapidly diminish.

An interesting example of the false alarm syndrome occurred in Australia in 1974 when the northern city of Darwin was virtually flattened by Cyclone Tracy. Although warnings of the cyclone's approach were timely, accurate, and clear, the general public paid inadequate regard to them and suffered accordingly. One of the main reasons for this was that only a short time previously a cyclone had tracked toward the city in an almost identical way, only to alter course prior to expected impact.

This question of warning overkill is obviously a difficult one for disaster managers to handle accurately. It does depend greatly on the efficacy of the warning sequence (paragraph 6 above) and the capability requirements outlined in paragraph 8. It also requires appropriate study by disaster managers and those responsible for initialising warnings, plus simulations of possible disaster scenarios.

## 17. *Training*

One wrongly transmitted or recorded message can disrupt the total warning process; and there is no doubt that expert and professional operation of all parts of the warning system must be a clear disaster management aim. Consequently, training requirements need to receive high priority. One advantage here is that training on the relevant communications procedures and information management is usually interesting for participants. Also, tests and simulations are not difficult to organize and implement. It should therefore be reasonably easy to meet the training requirement for the use of warning.

*Appendix C*

# Survey and Assessment

## Purpose

1. The purpose of this appendix is to outline the main considerations which apply to the survey and assessment of the after-effects of a disaster.

2. The appendix is primarily related to the response activities outlined in Chapter 17. It contains occasional reiteration from that chapter and this is intentional for purposes of emphasis.

3. The appendix covers the following main aspects:

- importance of survey and assessment,
- scope of survey and assessment,
- basis of planning,
- identification of information needs,
- use of existing information,
- nature of crisis information,
- systems and facilities,
- communication and coordination of information,
- warning lead time,
- post-impact action, and
- support arrangements.

## The Importance of Survey and Assessment

4. Survey and assessment is critical in responding to disaster because it covers two vital areas:

- First, it identifies what the particular disaster has done to the area concerned.
- Second, it indicates what needs to be done in terms of response (and later, recovery).

## The Scope of Survey and Assessment

5. In practice, survey and assessment may comprise a series of actions, the timing of which is largely determined by the particular disaster circumstance. However, the prime objectives of such actions are usually:

- To obtain a first general picture of the post-impact situation.
- To establish the state of accessibility into and within the stricken area for purposes of rescue, medical relief, supply, and other urgent action.
- To determine the needs of stricken communities in terms of immediate relief and long-term recovery.
- To determine the levels of damage which have occurred to buildings and structures, for purposes of restoration and reconstruction.
- To assess or verify crop losses which may need to be made good by seed or foodgrain imports.

6. Other survey and assessment action may be needed for specialist reasons, for instance:

- To establish the spread of contamination from hazardous substances (e.g., radiation toxic risks).
- To establish other threats which might affect the health of stricken communities.

## The Basis of Planning

7. The success of survey and assessment depends considerably on the basis of planning, organization, and preparedness measures. Survey and assessment action therefore needs to be carefully covered in disaster plans and any associated standard operating procedures. This is often done by:

- Allocating overall responsibility for coordinating survey and assessment to one authority (e.g., the police, because of their organizational spread and availability of communications; or local government authorities) and
- Allocating detailed areas of responsibility to various resource organizations (e.g., agriculture assessment to the Department of Agriculture; casualty and health assessment to medical and health agencies).

No matter how this aspect is dealt with, clear allocation of responsibility is most important. Otherwise, under the pressure of post-impact events, attempts to secure adequate and accurate information can become confused; and when this happens, overall response can be affected. There must therefore be a good basis of planning for survey and assessment.

## Identification of Information Needs

8. It is necessary to establish before disaster strikes what sort of information is likely to be needed. This can be done through a fairly simple process of definition, as follows:

- Define the likely threats.
- Define the effects likely to arise from these threats.
- Define what sort of information is needed to deal with these effects.

For example, information is usually needed on casualties, destruction of homes and property, damage to essential services, evacuation needs or possibilities, availability of food supplies, and so on. These areas of information can be readily established as items of preparedness.

The important point here, of course, is that before we are called upon to respond we need to know, quite simply and clearly, what we are looking for in terms of information.

9. This may sound an easy and simple matter. In fact, it is. The problem is that so often disaster management authorities make it difficult for themselves because they do not clearly identify their information needs in the planning and preparedness stage. On the other hand, where information needs have been clearly spelled out in disaster plans and procedures, this has paid handsome dividends during response action.

## The Use of Existing Information

10. It should be borne in mind that some information needs can be met from data that already exists. Thus, before survey and assessment is planned, it is necessary to identify what existing information may be used. Examples of such information may include:

- Information from previous disasters,
- Map information,
- Census information, and
- Specialist information of various kinds.

Such information needs to be of easy access when required.

## The Nature of Crisis Information

11. Crisis information, the information arising from the disaster area itself, must be:

- accurate,
- as up-to-date as possible, and
- sufficiently detailed to facilitate the organization of an accurate response.

Thus, taking the medical and health area as a brief summarized example, the overall purpose of survey action is likely to be:

- To ascertain the nature and extent of casualties.
- To assess the immediate medical and health needs of the stricken community.
- To assess surviving medical and health resources.

Detailed information needs would therefore include:

- categorization of casualties;
- state of the surviving medical and health system (including medical and health facilities, personnel, equipment supplies, medical transport, etc.);
- medical and health needs (including facilities, personnel, equipment, supplies, transport, etc.);
- medical evacuation needs;
- future risk factors (including sanitation problems, water supplies, overcrowding, nutritional problems; see also paragraph 6 above); and
- self-help and volunteer resources (e.g., personnel, improvised accommodation) from within the community.

12. Crisis information would also be required to cover other areas such as:

- essential services,
- food and agriculture,
- communications,
- transport, and
- housing and shelter.

## Systems and Facilities

13. The process suggested in paragraphs 8–12 could be expected to narrow down the situation so that it is clear what information needs to be sought. The next question, therefore, is what systems and facilities are usually available for acquiring and monitoring crisis information. These systems and facilities can be divided into two broad categories, as described below.

### 14. *Basic Systems*

*Emergency services*

Police, fire authorities, ambulance services, civil defense (or similar organizations).

*Government departments and agencies*

Those which have disaster-related roles.

For example, public works agencies are usually involved in assessing damage to roads, etc.; medical and health departments have a self-evident role in their specialist area; local government officials or teams have a role at that level.

*Nongovernment organizations*

NGOs can also provide useful support, as long as their responsibilities are clearly defined, agreed, and accepted.

*Media*

The media can also contribute information, though it is not always prepared to regard itself as a formal part of the counter-disaster system, even temporarily.

*General Public*

Reports from victims or observers in the disaster area are also useful.

*Special systems*

Satellite information (especially for movement of cyclones and major flooding and landslides after earthquake).

### 15. *Emergency Systems*

*Air survey*

Visual air surveys are usually valuable, especially for initial surveys. The "gyro-stabilized human eyeball" is an effective information-gathering mechanism. However, its owner does need to be experienced in air survey, otherwise inaccuracies can occur.

*Photographic* air surveys are also invaluable, particularly for obtaining a first assessment of the post-impact situation. Line overlaps and mosaics are very useful when the disaster is widespread.

*Ground Survey*

Ground surveys can include waterborne surveys of coastal areas and inland waterways. Their particular value is the collection of on-site data, including contact with the stricken community. These surveys usually need to be led by officials specialized in various fields.

*Air-grounds urvey*

In many cases, helicopter capability has proved invaluable for providing a form of urgent air-ground survey. This can be done by using a team of, say, an administrative official, a medical officer, and a disaster management official. A series of speedy visits carried out in this way is not only extremely valuable for providing an urgent initial assessment; it also helps assure stricken communities that assistance is under way.

Further information on initial surveys and follow-up surveys is contained in paragraphs 21–22 below.

## Communication and Co-ordination of Information

16. With regard to the communication and coordination of information, various problem areas can be identified from case study material. These include the following:

- amount and variety of information which may be generated by the disaster; some of this are often irrelevant and can confuse the assessment process;
- submission of information which is inaccurate; for example, from inexperienced observers;
- large number of transmission links or networks which may be involved; this may slow down the collation of information; and
- general effect of these and other factors which results in what is often called "information convergence" and which may choke the information management system, at least temporarily.

In practice, what this may mean is that it is not always possible to achieve the ideal situation whereby survey and assessment consists of a tidy, orderly process in which information is accurately collected, smoothly transmitted, and precisely coordinated.

It is important for disaster managers to be aware of this possibility, especially where disaster circumstances are severe.

### 17. *Standard Formats*

The use of standardized concise formats through the survey and assessment process can help to minimize the problems of communication and coordination. This is not always practicable but is worth working toward as an overall survey and assessment guideline (see also paragraphs 22–23).

### 18. *Communication Networks and Links*

There may some basic conflict in the use of communication networks and links because, in principle, it is not advisable to change normal day-to-day systems and procedures to meet disaster circumstances. At the same time, the information being transmitted over basic systems may, under disaster conditions, be required at coordinating points which are different from the normal ones. For example it may be necessary for certain emergency services to report information directly to, say, a mobile EOC.

The problems here are surmountable. They can be overcome by the use of emergency links and so on. However, these problems do need to be resolved during the planning process and not left until response operations are imminent or under way.

### 19. *Importance of Emergency Operations Centers*

The importance of effective EOCs in the process of survey and assessment cannot be overemphasized. Unfortunately, numerous cases are on record where this importance has been disregarded or overlooked. The results are usually very disruptive to the disaster-management process. For instance, survey information and other reports are often available from affected areas. However, this information cannot be put together without a reasonable EOC system. This results in inaccurate decision making, slow or inappropriate response action, and unnecessary additional hardship for people in the stricken area.

## Warning Lead Time

20. Warning lead time has an important bearing on effective survey and assessment. If good warning lead time of the type, timing, and likely severity of an oncoming disaster is available, this helps in terms of readiness to carry out survey and assessment. It also indicates the kind of post-impact surveys which may be needed.

## Post-Impact Action

### 21. *Initial Surveys*

Usually, the most important first action after disaster impact is to obtain an overall idea of the extent and severity of the effects. Initial surveys are therefore vitally important. In many circumstances, however, there tend to be constraints on these initial surveys. For example:

- Weather conditions following a cyclone (heavy rain and poor visibility) tend to hamper air survey.
- With an earthquake, extensive destruction and damage can make accurate assessment difficult.
- With floods also it is often difficult to make a quick, accurate assessment.

There also tends to be the general problem that, on the ground, access to and movement within most disaster areas can be difficult. This especially applies initially which is, of course, a critical period for information gathering.

Sometimes, international assistance can be invaluable for initial survey. Specialist reconnaissance aircraft can have a quick response and excellent facilities for photographic and visual survey. If it is possible (as is sometimes the case) for the aircraft to carry an official from the stricken country to help identify affected areas, this is advantageous. However, once again, this kind of action depends greatly on adequate preparedness.

With initial survey generally, wide use can made on both rotary and fixed wing aircraft. However, it does need to be remembered that visual air survey can be prone to inaccuracies, especially without factual back-up reports from the ground; and this can often be the case immediately post-impact.

## 22. *Follow-up Surveys*

The limitations of initial surveys obviously underline the importance of follow-up surveys. In most cases the only way of achieving these follow-up surveys is by hard, persistent work on the ground, which emphasizes the need for trained teams and a well-organized system.

A case study offers some useful pointers on these follow-up surveys. In this instance, a major survey became necessary because information from several earlier surveys was obviously incomplete. Particular points of interest were:

- First, a careful assessment was made of information from the earlier surveys. A lot of this earlier information had, incidentally, been gathered by international assistance resources during the early relief operations (when they delivered food and shelter materials, etc.)

- Second, the areas to be surveyed and the respective priorities were established.

- Third, careful definition was made of the additional information needed. This was not too difficult because ongoing relief operations provided a good guide to the important information categories.

- Fourth, resources to carry out the survey were assessed and use was made of various personnel groups. For example: including overseas emergency teams who had been flown in to help in the initial response operations; the local police, including their mobile force element; various government specialists; a variety of communications resources; teams of local inhabitants, under village leaders.

With these arrangements and resources, the survey was successfully completed and the necessary information obtained. However, it then became necessary to borrow a computer analyst from a neighboring country to make full use of all information collected. It is also of interest that, in this particular survey, the forms for recording information were specially drawn up beforehand and not taken from a standard manual or plan. This raises the point that there can be flexibility in the format used for surveys. The important basic requirements are to establish what information is being sought and then to follow a few commonsense guidelines and procedures.

## 23. *Development of Information Picture*

The progressive development of the information picture is absolutely crucial; and "picture" is the operative word. The coordinator or disaster executive does need a well-collated display from which to make critical response decisions. The exact form of that display will obviously depend on individual circumstances and facilities; but fairly simple status boards and an accurate map display will usually suffice. An important point worth remembering is that the information picture (apart from its obvious use in allocating response tasks) also helps the coordinator to decide what additional survey and assessment action needs to be taken.

## Support Arrangements

24. The foregoing suggested measures obviously need various support arrangements. For instance:

- *Training and Preparedness*
  This needs to cover aspects such as:
  - Adequate pre-briefing of those persons within the counter-disaster system who are likely to carry out survey and assessment tasks.
  - Designation of team leaders and deputies, as necessary.
  - The provision of guidelines and report forms (see the sample survey format at the end of this appendix).
  - Practice runs.

- *Administration and Logistics*
  For instance, transport and movement difficulties can often lead to delays in mounting both initial and follow-up surveys. Terrain can also cause logistic problems. These aspects, which must obviously be related to individual circumstances, need to be carefully considered and their effects anticipated as far as possible in guidelines and procedures.

## Summary

25. The information contained in this appendix is necessarily a fairly limited coverage of survey and assessment, primarily intended as a basis to be developed for individual requirements. In summary, the following are emphasised as key points:

- need for a sound basis of planning and preparedness;
- need to know clearly what information is being sought;

- need for astute utilization of available systems and facilities; and
- need to tackle the survey and assessment requirement, on a prepared and organized basis, because it will not wait, it will not go away, and people will suffer and die if survey and assessment is not effective.

*This sample survey format is reprinted with acknowledgment, from Post-Disaster Assessment by John R. Campbell and Joseph Chung sponsored by the Pacific Islands Development, East-West Center, Honolulu, Hawaii, USA. This manual was the outcome of a Workshop held at P1DP, of which the author of this handbook was a member.*

**Medical Survey**

Village/Settlement ______________________ Date ______________
Island/District. ______________________
*Population:* Adults (over 15 years)* ________
Children (5 to 15 years)* ________
Infants (below 5 years)* ________

*Medical Facilities*
(circle appropriate answer)
Clinic/Hospital/No Facility

Destroyed/
Damaged but Usable/
Undamaged
Beds Undamaged ____________
Beds Occupied ____________
Beds Available ____________

*Disease Risk Factors:*
Toilets:
Destroyed ____________
Damaged ____________
Serviceable ____________
Is there any problem with the disposal of sewage____________

*Medical Supplies:*
Lost ____________________
____________________
Remaining ____________________
____________________
Needed ____________________
____________________

*Medical Personnel* (by number):
Doctors Available ____________
Nurses Available ____________

*Other Risks:*
Stagnant Water ____________
Poor Drainage ____________
Poor Hygiene Conditions ________
Substandard Kitchens ________
Overcrowding ____________
Insect/Rat Infestation ________
Is Spraying Necessary ________

Remarks:
________________________________________
________________________________________

Name of Surveyor
________________________________________

* The ages given here are intended only as a guide. In some countries a different age breakdown is used to distinguish adults from children and children from infants.

*Appendix D*

# Examples of Disaster Legislation

## COOK ISLANDS

The legislation for the Cook Islands is given below.

### Analysis

*Title*

1. Short Title
2. Interpretation
3. Chief Hurricane Safety Officer
4. Hurricane Safety Officers
5. Functions and Powers of Chief Hurricane Safety Officer
6. Delegation of Powers of Chief Hurricane Safety Officer
7. Hurricane Safety Committee
8. Functions of Hurricane Safety Committee
9. Preparation and Approval of National Hurricane

*Safety Plans*

10. Obligation to conform to operative plan
11. State of Warning
12. Declaration of State of Emergency
13. Compensation for injury and loss of or damage to personal property
14. Protection from liability
15. Requisitioning powers
16. Evacuation of places and buildings
17. Obstruction or impersonation of officers
18. Penalty for offences
19. Regulations
20. Annual Report

### An Act to Provide for Hurricane Safety

18 May 1973

BE IT ENACTED by the Legislative Assembly of the Cook Islands in Session assembled, and by the authority of the same, as follows:

1. *Short Title* – This Act may be cited as the Hurricane Safety Act 1973
2. *Interpretation* – In this Act, unless the context otherwise requires:

"Minister" means the Minister charged with the responsibility of hurricane safety.

3. *Chief Hurricane Safety Officer*

(1) The Minister shall from time to time by notice in the Gazette appoint-

(a) A Chief Hurricane Safety Officer, stationed in Rarotonga who shall be charged with the duty of carrying this Act to effect; and

(b) A Deputy Chief Hurricane Safety Officer, stationed in Rarotonga, to act in case of the illness, absence, death or removal of the Chief Hurricane Safety Officer.

(2) The Deputy Chief Hurricane Safety Officer while so acting, shall exercise the duties and shall have the powers and authorities of the Chief Hurricane Safety Officer, and the fact that he so acts shall be sufficient evidence of his authority to do so.

4. *Hurricane Safety Officers*

(1) The Chief Hurricane Officer shall from time to time appoint–

(a) Hurricane Safety Officers, who will be in charge of Hurricane Safety arrangements for the island or district to which they are appointed:

(b) Supervisors, who will be in charge of Safety Centres and posts to which they are appointed:

(c) Hurricane Safety Officers for each village or tapere on the recommendation by the Village Committee for that area:

(d) Such other officers as in the opinion of the Chief Hurricane Safety Officer are required in any island to implement the provisions of this Act:

(e) A substitute for any officer appointed under this section to act in the case of the illness, absence, death or removal of that officer. The substitute while so acting, shall exercise the duties and have the powers and authorities of the officer for whom he is acting, and the fact that any substitute so acts shall be sufficient evidence of his authority to do so.

(2) Any reference in this Act to one of the officers appointed under this section shall be deemed to include any substitute appointed for that officer.

(3) Every person appointed under this Act shall exercise his duties and functions subject to the control of the Chief Hurricane Safety Officer; and shall comply with any directions received from him from time to time.

5. *Functions and powers of Chief Hurricane Safety Officer*

(1) The Functions of the Chief Hurricane Safety Officer shall be–

(a) To advise and assist the Minister in all matters relating to hurricane safety:

(b) To exercise and perform such functions, duties and powers with respect to hurricane safety as are conferred or imposed on him by this Act.

(2) Without limiting the generality of the provisions of subsection (1) of this section, the Chief Hurricane Safety Officer shall have the following specific functions:

(a) The co-ordination of the planning and execution generally of hurricane safety measures:

(b) The taking of all steps necessary to provide, or to arrange the provision of, or otherwise to render readily available, personnel, material, and services adequate for the effective carrying out of hurricane safety:

(c) The direction and control for the purposes of this Act of personnel, material, and services available and at his disposal for hurricane safety purpose.

(3) In the exercise of his functions, the Chief Hurricane Safety Officer may –

(a) Co-ordinate the use of, and during a state of warning or of emergency use, the personnel, material, and services made available by Departments of State, other Government agencies, statutory corporations, public bodies, other organizations, and otherwise for the purpose of hurricane safety, and in particular, without limiting the generality of the foregoing provisions of this paragraph, for-

(i) The provision of transport;

(ii) The removal of endangered persons and casualties from any area affected by hurricane emergency to areas of safety or to hospitals;

(iii) The affording of medical care and attention to casualties;

(iv) The relief of distress and suffering;

(v) The accommodation, feeding, care, and protection of persons;

(vi) The provision of other services necessary for the welfare of the public;

(vii) The purposes of this Act generally as may be directed by the Chief Hurricane Safety Officer;

(b) Enter into arrangements with any person whereby that person gives or undertakes to give his services or those of his staff, whether individually or in units under the control of that person, for the purpose of carrying out such hurricane safety measures as may be agreed upon:

(c) Devise, promote, and carry out, or cause to be carried out, research investigations into matters relating to hurricane safety, and promote and carry out, or cause to be carried out, the dissemination of information and advice on matters relating to hurricane safety:

(d) Promote and carry out, or cause to be carried out, the training of personnel for hurricane safety purposes.

(4) During a state of warning or emergency the Chief Hurricane Safety Officer shall have and may exercise, in addition to the functions and powers specified in this section, such functions and powers as may be conferred on him by regulations made under this Act.

6. *Delegation of powers of Chief Hurricane Safety Officer*

(1) The Chief Hurricane Safety Officer may from time to time, by writing under his hand, either generally or particularly, delegate to any employee of the Crown or any statutory corporation all or any of the powers exercisable by the Chief Hurricane Safety Officer under this Act, except this power of delegation.

(2) Subject to any general or special directions given or conditions attached by the Chief Hurricane Safety Officer, any person to whom any powers are delegated under this section may exercise those powers in the same manner and with the same effect as if they had been conferred on him directly by this section and not by delegation.

(3) Every person purporting to act pursuant to any delegation under this section shall be presumed to be acting in accordance with the terms of the delegation in the absence of truth to the contrary.

(4) Any delegation under this section may be made to a specific employee or to employees of a specified class, or may be made to the holder or holders for the time being of a specified office or class of offices.

(5) Every delegation under this section, shall be revocable in writing at will, and no such delegation shall prevent exercise of any power by the Chief Hurricane Safety Officer.

(6) Every delegation under this section shall, until revoked, continue in force according to its tenor. In the event of the Chief Hurricane Safety Officer by whom any such delegation has been made ceasing to hold office, it shall continue to have effect as if made by the person for the time being holding office as Chief Hurricane Safety Officer.

7. *Hurricane Safety Committee*

(1) For the purposes of this Act there shall be a Committee, to be called the Hurricane Safety Committee.

(2) The Committee shall consist of –

(a) The Chief Hurricane Safety Officer, who shall be Chairman;
(b) The Superintendent of Police;
(c) The Director of Works;
(d) The Superintendent of Radio;
(e) The Director of Health;
(f) Chief Postmaster;
(g) Secretary of Internal Affairs;
(h) Such other persons as are appointed by the Chief Hurricane Safety Officer with the approval of the Minister.

8. *Functions of Hurricane Safety Committee –*

(1) The general functions of the Hurricane Safety Committee shall be to advise and assist the Chief Hurricane Safety Officer in planning and implementing all measures considered necessary or desirable for the establishment, maintenance, and effective operation of hurricane safety.

(2) The Committee shall determine its own procedure.

9. *Preparation and approval of National Hurricane Safety plans –*

(1) Not later than the first day of November in each year the Chief Hurricane Safety Officer shall forward to the Minister a national plan in respect of hurricane safety for the immediately ensuing hurricane season.

(2) Where the Minister is satisfied that any national plan prepared under this section provides adequately for hurricane safety measures, he may approve the plan, and the plan shall thereupon be deemed to be operative.

10. *Obligation to conform to operative plan* – Every department of state, other government agency, or statutory corporation required or authorised by any operative plan, or regulations made under this Act to undertake any hurricane safety measures or to exercise any functions shall take all necessary steps to undertake those measures or to exercise those functions.

11. *State of Warning –*

(1) A state of warning shall exist from the time a preliminary warning is publicly notified by such means of communication as are available by the Chief Hurricane Safety Officer, or by the New Zealand Meteorological Service.

(2) The state of warning shall continue until an all clear signal is publicly notified by such means of communication as are available by the Chief Hurricane Safety Officer or the New Zealand Meteorological Service.

12. *Declaration of state of emergency –*

(1) if at any time it appears to the Minister that –

(a) A civil emergency has occurred or may occur; and

(b) The emergency is or is likely to be of such extent or magnitude or severity that national measures are necessary or desirable in respect of it, he may declare a state of emergency for any island or islands, as he sees fit.

(2) The Minister shall forthwith give public notice, by such means of communications as are available, of every declaration of a state of emergency made by him under subsection (1) of this section; and the declaration shall be gazetted as soon as practicable.

(3) A state of emergency shall terminate on the seventh day after the date on which it was declared.

13. *Compensation for injury and loss of or damage to personal property –*

(1) Every person who while carrying out hurricane safety work, or participating in hurricane safety training, under the control of any person appointed under sections 3 or 4 of this Act, suffers bodily injury or loss or damage to his personal property, shall be entitled to receive compensation equal to –

(a) All expenses directly incurred as a result of the injury; or

(b) The value of any personal property that has been lost; or

(c) The diminution in value of any personal property that has been damaged – as the case may be, if injury or loss or damage was directly attributable to carrying out such work or the participation in such training.

(2) Any compensation payable under subsection (1) of this section shall be paid by the Crown out of moneys appropriated by the Legislative Assembly for the purpose.

(3) In the event of any dispute as to the proper compensation payable under this section the amount of compensation payable be fixed by the High Court.

(4) For the purposes of this section, travelling directly to and returning directly from any place where hurricane safety training is to be or has been carried out shall be deemed to be the carrying out of such work or the participation in such training as the case may be.

14. *Protection from liability –*

(1) No action or proceedings shall be brought against the Crown, or the Chief Hurricane Safety Officer, or any officer or servant of them, or against any other person whatsoever to recover damages for any damage to property occasioned by any person in the exercise or performance in good faith of his powers, duties, or obligations under this Act.

(2) No person shall be personally liable for any act done or default made by him in good faith in the course of carrying out hurricane safety work, or participating in hurricane safety training, under the control of a person referred to in sections 3 or 4 of this Act.

15. *Requisitioning powers –*

(1) Where a state of warning exists or a state of emergency has been declared, and, in the opinion of the Chief Hurricane Safety Officer or member of the Police, the action authorised by this subsection is immediately and urgently necessary for the preservation of human life, he or any person authorised by him may, without any further authority than this subsection, require the owner or person for the time being in control of any vehicle, horse, boat, apparatus, implement, earth-moving equipment, construction equipment, or other equipment of any kind whatsoever (hereinafter in this section referred to as requisitioned property) forthwith to place that requisitioned property under the control and direction of the Chief Hurricane Safety Officer, Hurricane Safety Officer or member of the Police, or person so authorised, as the case may be:

Provided that where the owner or person for the time being in control of that requisitioned property cannot be immediately found the Chief Hurricane Safety Officer, Hurricane Safety Officer or member of the Police, or person so authorised may assume forthwith the control and direction of the requisitioned property.

(2) Every person commits an offense against this Act who fails to comply with any direction given to him under subsection (1) of this section.

Provided that no person shall be deemed to have committed an offence under this section unless the Court is satisfied that the Chief Hurricane Safety Officer, Hurricane Safety Officer or member of the Police or person so authorised had reasonable grounds for believing that in all circumstances of the case the requirement was immediately and urgently necessary for the preservation of human life.

(3) Where under this section any requisitioned property has come under the control of a Chief Hurricane Safety Officer, Hurricane Safety Officer or member of the Police, there shall, on application by the owner, be paid to him out of money appropriated by Legislative Assembly such amount as the Minister approves by way of reasonable compensation for –

(a) The use of that requisitioned property:

(b) Any loss or damage or injury to that requisitioned property suffered or incurred while under that control.

16. *Evacuation of places and buildings* –

(1) Where a state of warning exists or a state of emergency has been declared, and, in the opinion of the Chief Hurricane Safety Officer or member of the Police, the action authorised by this subsection is immediately and urgently necessary for the preservation of human life, he or any person authorised by him may, without any further authority than this subsection, require the evacuation of any area or building or place and the exclusion of any persons and vehicles from any area, building, or place.

(2) Every person commits an offence against this Act who fails to comply with any direction given to him under subsection (1) of this section:

Provided that no person shall be deemed to have committed an offence under this section unless the Court is satisfied that the Chief Hurricane Safety Officer, Hurricane Safety Officer or member of the Police or person so authorised had reasonable grounds for believing that in all circumstances of the case the requirement was immediately and urgently necessary for the preservation of human life.

17. *Obstruction or impersonation of officers* – Every person commits an offence who willfully –

(a) Prevents the Chief Hurricane Safety Officer, any person acting under the authority of the Chief Hurricane Safety Officer, or any member of the Police, or any person duly authorised or employed for the purpose of carrying out any provision of or under this Act or of any such provision; or

(b) Obstructs or impedes any such person in carrying out any such provision; or
(c) Not being a person referred to in paragraph (a) of this section, for the purpose of committing or facilitating the commission of an offence under any enactment, by words, conduct, or demeanour pretends that he is such person, or puts on or assumes the dress, name, designation, or description of such person.

18. *Penalty for offences* – Every person who commits an offence against this Act is liable on conviction to imprisonment for a term not exceeding four hundred dollars.

19. *Regulations* –
(1) The High Commissioner may from time to time, by Order in Executive Council, make such regulations with respect to hurricane safety as appear to him to be necessary or expedient for the purpose of securing the public safety and generally safeguarding the interests of the public during any warning or emergency.

(2) Regulations made under subsection (1) of this section may describe penalties for the breach of any such regulation, not exceeding –
(a) In the case of an offence committed willfully by an individual, imprisonment for a term of three months or a fine of two hundred dollars:
(b) In the case of any other offence committed by an individual, a fine of one hundred dollars.

(3) The High Commissioner may from time to time, by Order in Executive Council, make regulations generally providing for such matters as are contemplated by or necessary for giving full effect to the provisions of this Act or any operative national plan and for the due administration of this Act.

(4) All regulations made under this Act shall be laid before the Legislative Assembly within twenty-eight days after the making thereof if the Legislative Assembly is then in Session, and, if not, shall be laid before the Legislative Assembly within twenty-eight days after the commencement of the next ensuing session.

20. *Annual Report* –
(1) The Chief Hurricane Safety Officer shall, no later than the thirty-first day of May in each year furnish to the Minister a report on the administration of this Act.

(2) A copy of the report shall be laid before the Legislative Assembly within twenty-eight days after it has been received by the Minister if the Legislative Assembly is in Session, and, if not, shall be laid before the Legislative Assembly within twenty-eight days after the date of Commencement of the next ensuing session.

This Act is administered by the Chief Hurricane Safety Officer.

RAROTONGA, COOK ISLANDS: Printed under the authority of the Cook Islands Government, by T. KAPI, Government Printer 1973.

# INDEPENDENT STATE OF PAPUA NEW GUINEA

## CHAPTER NO. 403

*Disaster Management Act*

---

ARRANGEMENTS OF SECTIONS

**PART I – PRELIMINARY**

1. Purpose of the Act.
2. Interpretation –
   "disaster"
   "the National Disaster Centre"
   "the National Disaster Committee"
   "province"
   "Provincial Disaster Committee"
   "this Act"

**PART II – DISASTER MANAGEMENT**

*Division 1 – National Disaster Committee*

3. National Disaster Committee.
4. Co-opted members of National Disaster Committee.
5. Meetings of the National Disaster Committee.
6. Functions of the National Disaster Committee.
7. Powers of the National Disaster Committee.
8. Reports.

*Division 2 – Provincial Disaster Committee*

9. Provincial Disaster Committees.
10. Co-opted members of Provincial Disaster Committees.
11. Meetings of a Provincial Disaster Committee.
12. Functions of a Provincial Disaster Committee.
13. Powers of a Provincial Disaster Committee.
14. Reports.

---

# INDEPENDENT STATE OF PAPUA NEW GUINEA

## CHAPTER NO. 403

Disaster Management Act

Being an Act to make provision with respect to emergencies arising out of epidemic, earthquakes, volcanic eruptions and other disasters, and for related purposes.

### PART I – PRELIMINARY

**1. Purpose of the Act.**

The purpose of this Act is to establish a machinery for forward planning to ensure the efficient, prompt and effective management and control of natural disasters in the country.

**2. Interpretation.**

In this Act, unless the contrary intention appears –

"disaster" means an earthquake, volcanic eruption, storm, tempest, flood, fire or outbreak of pestilence or infectious disease, or any other natural calamity whether similar to any such occurrence or not, on such an extensive scale as to be likely to endanger the public safety or to deprive the community of supplies or services essential to life;

"the National Disaster Centre" means the National Disaster Centre established by Section 15;

"the National Disaster Committee" means the Committee established by Section 3;

"province" includes the National Capital District;

"Provincial Disaster Committee" means the Committee established by Section 9;

"this Act" includes the regulations.

## PART II – DISASTER COMMITTEE

*Division 1 – National Disaster Committee.*

### 3. National Disaster Committee.

(1) The National Disaster Committee is hereby established.

(2) The National Disaster Committee shall consist of –

(a) the Secretary to the Department of the Prime Minister; and
(b) the Commander of the Defence Force; and
(c) the Commissioner of Police; and
(d) the Head of the Department responsible for financial matters; and
(e) The Head of the Department responsible for defence matters; and
(f) the Head of the Department responsible for works and supply matters; and
(g) the Head of the Department responsible for health matters; and
(h) where matters concerning international assistance are under consideration, the Head of the Department responsible for foreign affairs and trade matters.

(3) The Prime Minister shall appoint one of the members of the National Disaster Committee to be the Chairman and another to be the Deputy Chairman.

### 4. Co-opted members of National Disaster Committee.

(1) The National Disaster Committee may, in relation to a matter before it or to a particular disaster, co-opt –

(a) any Department Head or officer; or
(b) any other person,

as a member of the National Disaster Committee.

(2) A member co-opted under Subsection (1) may take part in deliberations and vote on questions relating to the matter or to the particular disaster in relation to which he was appointed.

(3) For the purposes of this Act, other than as provided by this section, a member co-opted under Subsection (1) shall not be regarded as a member of the National Disaster Committee.

5. **Meetings of the National Disaster Committee.**

(1) The National Disaster Committee may meet and transact business at such times and in such places as determined by the Chairman, and in any case shall meet at least twice in a calendar year.

(2) Where he receives a request to do so by a member of the National Disaster Committee, the Chairman shall convene a meeting within three days.

(3) At a meeting of the National Disaster Committee –
(a) a quorum is three members; and
(b) the Chairman shall preside at all meetings at which he is present; and
(c) in the absence of the Chairman, the Deputy Chairman shall preside and where both the Chairman and Deputy Chairman are absent, the members present shall elect one of their number to act as Chairman for the purposes of that meeting; and
(d) matters arising shall be determined by a majority of votes of members present and voting; and
(e) the person presiding has a deliberate and, in the event of an equality of votes on a matter, also a casting vote.

(4) The National Disaster Committee shall cause full minutes of its meetings to be kept in such a manner as determined by it.

(5) Except where a specific disaster is being considered, a member of the National Disaster Committee may be represented at a meeting of that Committee by a delegate appointed in writing by the member and a delegate shall have the full powers of the member for whom he is delegate.

(6) Subject to this section, the procedures at a meeting of the National Disaster Committee are as determined by it.

6. **Functions of the National Disaster Committee.**

The functions of the National Disaster Committee are –

(a) to supervise the national state of preparedness for emergencies and to report on it to the National Executive Council; and

(b) to maintain the National Emergency Plan of the National Executive Council; and

(c) to assign responsibilities for disaster-related activities to Department and other bodies; and

(d) to co-ordinate Departmental relief actions and collate national relief requirements; and

(e) to advise the National Executive Council if appeal for international assistance is required; and

(f) to lay down guidelines for the preparation and format of provincial disaster plans; and

(g) to examine and where necessary approve all grants from national funds for emergency relief; and

(h) to foster public awareness of the effects of natural phenomena and measures which can be taken to prevent or mitigate them; and

(i) to supervise the establishment of stock-piles of relief supplies; and

(j) in the case of a disaster, to advise the National Executive Council of all the circumstances and on the advisability of declaring a National Emergency under the Constitution.

7. **Powers of the National Disaster Committee.**

The National Disaster Committee has power to do all things necessary or convenient to be done for or in connection with the performance of its functions under this Act.

8. **Reports.**

(1) The National Disaster Committee shall, as soon as possible after 31 December in each year, prepare and give to the Minister for presentation to the National Parliament, a report on the activities of the Committee for the year ending 31 December.

(2) Subsection (1) does not prevent the National Disaster Committee from making, on its own initiative, or at the request of the Minister, other reports on the work of the National Disaster Committee.

*Division 2 – Provincial Disaster Committees*

**9. Provincial Disaster Committee**

(1) A Provincial Disaster Committee is hereby established for each province.

(2) A Provincial Disaster Committee shall consist of –

(a) the head of the Department of the Province or, in the case of the National Capital District, the Manager of the National Capital District Interim Commission, who is the Chairman; and

(b) the Provincial Police Commander or, in the case of the National Capital District, the National Capital District Police Commander; and

(c) the Provincial Manager (Works and Supply) or, in the case of the National Capital District, the National Capital District City Engineer; and

(d) the Provincial Health Officer or, in the case of the National Capital District, the National Capital District Health Officer; and

(e) the officer-in-charge of the Provincial Affairs Branch of the Department of the Province or, in the case of the National Capital District, a person appointed by the Manager of the National Capital District Interim Commission.

**10. Co-opted members of the Provincial Disaster Committees.**

(1) A Provincial Disaster Committee may, in relation to a matter before it, or to a particular disaster, co-opt –

(a) an officer; or

(b) a member of a mission or voluntary organization; or

(c) any person,

as a member of the Provincial Disaster Committee.

(2) A member co-opted under Subsection (1) may take part in deliberations and vote on questions relating to the matter or to the particular disaster in relation to which he was appointed.

(3) For the purposes of this Act, other than as provided in this section, a member co-opted under Subsection (1) shall not be regarded as a member of a Provincial Disaster Committee.

**11. Meetings of a Provincial Disaster Committee.**

(1) A Provincial Disaster Committee may meet and transact business at such times and in such places as determined by the Chairman.

(2) At a meeting of a Provincial Disaster Committee –
- (a) a quorum if three members; and
- (b) the Chairman shall preside at all meetings at which he is present, and, in the absence of the Chairman, members present shall elect one of their number to preside; and
- (c) matters arising shall be decided by a majority of votes of members present and voting; and
- (d) the person presiding has a deliberative and, in the event of an equality of votes on a matter, also a casting vote.

(3) A Provincial Disaster Committee shall cause full minutes of its meetings to be kept in such a manner as determined by it.

(4) Subject to this section, the procedures at a meeting of a Provincial Disaster Committee shall be as determined by it.

**12. Functions of a Provincial Disaster Committee.**

The functions of a Provincial Disaster Committee are –
- (a) to assess particular hazards facing the province; and
- (b) to liaise and co-operate with the Provincial Executive Council in ensuring that development plans for the province take into account hazards facing it; and
- (c) to prepare emergency plans for the province in accordance with guidelines laid down by the National Disaster Committee, and supervise the state of preparedness for emergencies in the province; and
- (d) to co-ordinate stock piling of relief supplies and relief operations; and
- (e) to receive all applications for relief assistance; and
- (f) to foster public awareness of the effects of natural phenomena and the measures which can be taken to prevent or mitigate them; and
- (g) to organise the training of relief workers and to practise the provincial emergency plans.

**13. Powers of a Provincial Disaster Committee.**

The Provincial Disaster Committee has powers to do all things necessary or convenient to be done for or in connection with the performance of its functions under this Act.

**14. Reports.**

(1) The Provincial Disaster Committee shall, as soon as possible after 31 December in each year, prepare and give to –

(a) the Premier, for presentation to the Provincial Assembly; and

(b) the Minister, for presentation to the Parliament, a report on the activities of the Committee for the year ending 31 December.

(2) A report under Subsection (1) shall include details of –

(a) emergency and other plans; and

(b) actions taken to implement the requirements of preparedness dealt within the plans referred to in paragraph (a); and

(c) all aid requests to, and grants from, the National Disaster Committee.

(3) Where necessary, the Provincial Disaster Committee shall furnish a report for the purposes of Section 19(2), and such report shall include a minority report (if any).

(4) This section does not prevent a Provincial Disaster Committee from making, on its own initiative, or at the request of the Minister, other reports on its activities.

## PART III – NATIONAL DISASTER CENTRE

**15. National Disaster Centre.**

(1) The National Disaster Centre is hereby established.

(2) The National Disaster Centre shall, in addition to such functions and duties as are provided under any other Act, co-ordinate all disaster situations and surveillance matters.

**16. Director of the National Disaster Centre.**

(1) There shall be a Director of the National Disaster Centre who shall be –

(a) an officer of a State Service; and

(b) appointed by the Minister, after consultation with the National Disaster Committee; and

(c) appointed for a term of not less than two years.

(2) The Director of the National Disaster Centre shall be the Executive Officer of the National Disaster Committee.

**17. Functions of the Director.**

The functions of the Director of the National Disaster Centre are –

(a) to manage the National Disaster Centre; and

(b) to provide assistance to the Provincial Disaster Committees; and

(c) such other functions and duties as are, from time to time, directed by the National Disaster Committee.

**18. Other staff.**

Any staff required for the purposes of the National Disaster Centre shall be members of the Defense Force.

## PART IV – FINANCIAL AND OTHER ASSISTANCE

**19. Request for assistance.**

(1) A Provincial Disaster Committee may request the National Disaster Committee for financial and other assistance.

(2) A request under Subsection (1) shall be accompanied by a report of the extent of the funding required.

(3) Subject to Subsection (4), the National Disaster Committee may, on receipt of a request and the report referred to in Subsections (1) and (2), direct the Director of the National Disaster Centre to investigate the request for assistance and to report to it.

(4) The requirement under Subsection (3) may be waived or varied where the National Disaster Committee considers the request to be a matter of urgency.

**20. Granting request for assistance.**

(1) Subject to Subsection (2), funding of disaster relief operations shall be in accordance with the following: –

(a) the first K15,000.00 shall be borne by the Provincial Government concerned; and

(b) any excess over K15,000.00 and up to K100,000.00 shall be borne equally between the National Government and the Provincial Government concerned; and

(c) any excess over K100,000.00 shall be borne by the National Government.

(2) Allamounts granted by the National Government for the purpose of Subsection (l)(*b*) and (*c*) shall be subject to such conditions as are imposed by the National Government.

## PART V – MISCELLANEOUS

**22. Regulations.**

The Head of State, acting on advice, may make regulations not inconsistent with this Act, prescribing all matters that by this Act are required or permitted to be prescribed or are necessary or convenient to be prescribed for carrying out or giving effect to this Act.

**23. Repeal.**

The *Emergency Provisions Act* is repealed.

# QUEENSLAND, AUSTRALIA

---

## STATE COUNTER-DISASTER ORGANIZATION ACT 1975-1978

[Reprinted as of 1 May 1981]

---

State Counter-Disaster Organization Act 1975, No. 40

As amended by

State Counter-Disaster Organization Act Amendment Act 1978, No. 60

---

An Act to provide for the establishment of a State Counter-Disaster Organization and a State Emergency Service and their powers, authorities, functions and duties and for matters incidental to and consequent upon their establishment

[Assented to 1 October 1975]

BE IT ENACTED by the Queen's Most Excellent Majesty, by and with the advice and consent of the Legislative Assembly of Queensland in Parliament assembled, and by the authority of the same, as follows: –

### PART I – PRELIMINARY

**1. Short title.**This Act may be cited as the *State Counter-Disaster Organization Act* 1975.

Collective title conferred by Act of 1978, No. 60, s. 1 (3).

**2. Commencement.**This act shall commence on a date appointed by Proclamation.

Commenced 11 December 1975 (Proc. pubd. Gaz. 11 December 1975, p. 1387).

3. **Arrangement.** This Act is divided into. Parts as follows: –

PART I – PRELIMINARY (ss. 1-7);

PART II – ADMINISTRATION (ss. 8-22);

PART III – DECLARATION OF STATE DISASTER (ss. 23-25);

PART IV – POWERS AND DUTIES OF LOCAL AUTHORITIES (ss. 26-28);

PART V – MISCELLANEOUS (ss. 29-38).

4. **Repeal and savings.**

(1) *The Civil Defense Acts*, 1939-1942 (in this Act referred to as the "repealed Acts") are repealed.

(2) A body by whatever name called established pursuant to the repealed Acts by a Local Authority or Combined Local Authorities and in existence at the commencement of this Act is continued in existence and shall be deemed to be established under this Act.

(3) Where a provision of this Act is inconsistent with a provision of any other Act, the provision of this Act shall prevail.

5. **Limitation of Operation.** This Act does not authorize the taking of measures amounting to or making preparations for –

(a) actual combat against an enemy;

(b) the putting down of a riot or other civil disturbance;

(c) the bringing to an end of a strike or lock-out.

6. **Interpretation.** In this Act, save where the contrary intention appears –

"authorized person" means a person authorized to act by the chairman or by a disaster district co-ordinator;

"body" means a body corporate or unincorporate and includes a government department, instrumentality or agency, public or Local Authority;

"Chairman" means the Chairman of the Central Control Group established under this Act;

"Combined Local Authorities" means the Local Authorities that have united with one another for counter-disaster purpose;

"counter-disaster" means the planning, organizing, co-ordination or implementation of measures that are necessary or desirable to prevent, minimize or overcome the effects of a disaster upon members of the public or any property in the State and includes the conduct of or participation in training for those purposes; the term includes also civil defence measures necessary to combat the effects of enemy attack or hostilities;

"Director" means the Director of the State Emergency Service constituted under this Act and includes the person for the time being carrying out the duties of the Director;

"disaster" means –

(a) a flood, earthquake, seismic sea wave, cyclone, storm, tornado, eruption or other natural happening;

(b) any explosion, fire, oil spill or accident of any kind;

(c) any infestation, plague or epidemic;

(d) an attack directed against the State whether or not made by an enemy and whether by bombs or missiles or by atomic, chemical or any means, that causes or threatens to cause loss of life property or injury to persons or property distress to persons or that in any way endangers the safety of the public in the State or any part thereof;

"disaster district" means a portion of the State declared to be a disaster district under this Act;

"disaster district co-ordinator" means a disaster district co-ordinator appointed under this Act;

"Local Authority" means a Local Authority within the meaning of the *Local Government Act* 1936-1975 and with respect to the City of Brisbane means Brisbane City Council constituted by the City *of Brisbane Act* 1924-1974;

"local controller" means the controller of a local emergency service appointed under this Act;

"local emergency service" means an emergency service established and maintained under this Act by a Local Authority within their combined Areas;

"Minister" means the Minister for Police or other Minister of the Crown who, at the material time, is charged with the administration of this Act and includes any Minister of the Crown who is temporarily performing the duties of the Minister;

"Organization" means the State Counter-Disaster Organization constituted under this Act;

"region" means any part of a disaster district or any disaster district or any two or more disaster districts declared to be a region under this Act;

"regional operations officer" means a regional operations officer appointed under this Act;

"resources" includes food, manpower, any horse or other animal, vehicle, vessel, aircraft, plant, apparatus, implement, earth-moving equipment, construction equipment of any kind or any means of supplying a want or need;

"statutory service" mean a body that is constituted by or under an Act of the State or of the Commonwealth and whose role usually includes counter-disaster operations;

"vehicle" means a vehicle within the meaning of the *Traffic Act* 1949-1977;

"vessel" includes ship, boat, punt, ferry or air cushion vehicle however propelled.

As amended by Act of 1978, No. 60, s. 2.

7. **Crown bound.** This Act binds the Crown.

## PART II – ADMINISTRATION

8. **Administration of Act.** This Act shall be administered by the Minister and subject to his control and direction by the Director and other officers appointed pursuant to this Act.

9. **State Counter-Disaster Organization.**

(1) There is established and there shall at all times hereafter be constituted a body called the "State Counter-Disaster Organization".

(2) The Organization shall consist of –

(a) the holders for the time being of the following offices of the State –

(i) the Co-ordinator-General;
(ii) the Director-General of Health;
(iii) the Under Treasurer;
(iv) the Undersecretary, Premier's department;
(v) the Commissioner of Police;
(vi) the Director of Harbours and Marine;
(vii) the Director, State Emergency Service;

(b) such other persons are appointed by the Governor in Council by Order in Council.

Persons appointed pursuant to subparagraph (b) shall hold office during the pleasure of the Governor in Council.

10. **Functions of Organization.** The function of the Organization are –

(a) to co-ordinate the resources necessary to ensure that all steps are taken to plan for and counter the effects of a disaster;

(b) to give advice and assistance to the Minister on all matters with respect to counter-disaster.

**11. Business of Organization.** The organization shall meet as often as necessary for the administration of this Act and the performance of its functions at such times and places as it thinks fit and shall conduct its business in the manner prescribed or, so far as not prescribed, as it determines from time to time.

**12. Central Control Group.**

(1) For the purposes of carrying out the functions of the Organization there is established and there shall at all times hereafter be constituted a Central Control Group consisting of those members of the Organization who are holders for the time being of those offices of the State specified in Section 9 (2) (a). The Co-ordinator-General shall be the Chairman of the Group and the Director, State Emergency Service shall be the executive officer.

(2) The Minister may, in consultation with the Minister in charge of any department of the Government of the State or statutory corporation, appoint as a member of the Central Control Group for such period as the Minister thinks fit the permanent head of that department or statutory corporation, whether or not that permanent head is a member of the Organization.

(3) The Chairman shall preside at all meetings or consultations among members of the Central Control Group at which he is present and in his absence from any cause another member thereof appointed as prescribed shall preside and while so presiding shall have the powers, authorities, functions, duties and immunities of the Chairman.

(4) If any member is unable from any cause to attend a meeting or consultation among members of the Central Control Group, he may authorize any other officer of his department to attend the meeting or participate in the consultation in his stead and while so attending or participating that other officer shall be deemed for all purposes to be a member of the Central Control Group save that he shall not be entitled to be chairman of that meeting or for the purposes of that consultation.

(5) The Central Control Group shall meet as often as necessary for the due performance of its functions at such times and places as it thinks fit and shall conduct its business in the manner prescribed or so far as not prescribed, as it determines from time to time.

**13. State Emergency Service.**

(1) The body called the "Queensland State Emergency Service" established pursuant to the repealed Acts is continued in existence and established under this Act under the name "State Emergency Service".

(2) The persons who at the commencement of this Act comprised the body called the Queensland State Emergency Service shall, subject to this Act, comprise the State Emergency Service as established by this Act.

(3) The Governor in Council may appoint a Director of the State Emergency Service and such other officers as he considers necessary for the effectual administration of the State Emergency Service.

An officer appointed pursuant to this subsection shall hold office under, subject to and in accordance with *Public Service Act* 1922-1973.

(4) Unless and until another appointment is made, the person who at the commencement of this Act is the Director of the body called the Queensland State Emergency Service shall be the Director and any other person holding an office or occupying a position in that body at the commencement of this Act shall continue to hold that office or occupy that position until he vacates or is lawfully removed from that office or position.

**14. Functions of State Emergency Service.** The functions of the State Emergency Service are –

(a) to advise and assist Local Authorities, Government Departments, statutory organizations, voluntary groups and other bodies;

(b) to educate and train members of the public (including volunteers and members of voluntary groups);

(c) to co-ordinate, direct and control members of the public (including volunteers and members of voluntary groups), materials and resources,

with respect to counter-disaster purposes.

Substituted by Act of 1978, No. 60, s. 3.

**15. Powers and duties of Director generally.** In carrying out the functions of the State Emergency Service, the Director –

(a) shall establish and direct the policy of the State Emergency Service with respect to planning, organization, equipment, training, administration and operations;

(b) shall arrange counter-disaster education and advisory programmes and disseminate information;

(c) may appoint suitable persons to be registered volunteer members of the State Emergency Service;
(d) may appoint suitable persons to be executive volunteer members (other than controllers) of the State Emergency Service;
(e) prior to, during or subsequent to occurrence of a disaster, may appoint suitable persons to be temporary volunteer members of the State Emergency Service;
(f) may issue to members or volunteer members of any class or kind adequate means of personal identification;
(g) may inspect at regular intervals of time resources provided for Local Authorities for counter-disaster purposes;
(h) shall advise and assist the Minister and the Chairman on all matters with respect to the State Emergency Service.

As amended by Act of 1978, No. 60, s. 4.

**16. Powers and duties of Director as executive officer.** In his capacity as executive officer of the Central Control Group, the Director –

(a) shall assist and advise the Minister and the Chairman on all matters with respect to counter-disaster;
(b) shall be responsible to the Chairman for the co-ordination and adequacy of counter-disaster measures;
(c) shall, prior to, during or subsequent to the occurrence of a disaster translate to action instructions the decisions of the Central Control Group and shall ensure that those instructions are transmitted to and carried out by the bodies to whom they are directed.

**17. Committees.** The Minister may appoint such executive committees and advisory committees as he considers necessary to assist the Organization of the State Emergency Service in the efficient performance of its functions or in the achievement of its objects and purposes.

A committee shall comprise prescribed persons.

**18. Declaration of regions.**The Governor in Council may, by Order in Council –

(a) divide a disaster district into two or more parts and declare each part to be a region for the purposes of this Act;
(b) declare a disaster district or any two or more disaster districts to be a region for the purposes of this Act;
(c) vary any region declared pursuant to paragraph (b) by excluding therefrom or including therein any disaster district.

Substituted by Act of 1978, No. 60, s. 5.

**19. Appointment of regional operations officer.**

(1) The Governor in Council may by Order in Council appoint in respect of each region a regional operations officer.

(2) A regional operations officer shall have and may exercise such powers and perform functions and duties as per prescribed or so as far as not prescribed as the Director determines.

**20. Disaster districts.**

(1) The Governor in Council may by Order in Council constitute any portion of the State a disaster district for the purposes of this Act and assign to that district a name and may in like manner –

(a) include in a disaster district any portion of the State that is not included in a disaster district;
(b) abolish a disaster district or districts and join such district or districts or join parts of such districts with another district or districts;
(c) alter the boundaries of disaster districts by including in one district any part or parts of other districts and by excluding such part or parts from such other district or districts;
(d) divide a disaster district into two or more disaster districts;
(e) alter the name of a disaster district.

(2) A disaster district constituted pursuant to this section may comprise the whole or any division or part of the Area of a Local Authority or the whole or any divisions or parts of the Areas of two or more Local Authorities.

**21. Disaster district control groups.**

(1) There is established and there shall at all times hereafter be constituted in respect of each disaster district a disaster district control group comprising a disaster district co-ordinator appointed by the Governor in Council by Order in Council who shall be chairman and such other members as are prescribed, appointed as prescribed.

(2) The functions of a disaster district control group are –

(a) to make counter-disaster plans for its disaster district and review them from time to time and submit plans and reviewed plans to the Central Control Group;
(b) to keep and maintain up to date standing orders for counter-disaster purposes within its disaster district;
(c) such other functions as are prescribed.

(3) A disaster district control group shall meet as often as is necessary for the due performance of its functions at such times and places as it thinks fit and shall conduct its business in the manner prescribed or, so far as not prescribed, as it determines from time to time.

(4) A disaster district co-ordinator shall have and may exercise such powers and perform such functions and duties as are prescribed or so far as not prescribed as the Minister determines.
As amended by Act of 1978, No. 60, s. 6

**22. Power to delegate.**
(1) The Minister, Chairman or Director may either generally or otherwise as provided by the instrument of delegation, by writing signed by him, delegate to any person all or any of his powers, authorities, functions and duties under this Act except this power of delegation.

(2) A power, authority, function or duty so delegated, if exercised or performed by the delegate; shall be exercised or performed in accordance with the instrument of delegation.

(3) A delegation may be made subject to such terms or limitations as the Minister, Chairman or Director thinks fit including a requirement that the delegate shall report to him upon the exercise or performance of the delegated power, authority, function or duty.

(4) The Minister, Chairman or Director may make such and so many delegations of the same power, authority, function or duty and to such number of persons as he considers necessary or desirable.

(5) A delegation is revocable at the will of the Minister, Chairman or Director and does not prevent the exercise of a power or authority or the performance of a function or duty by him.

## PART III – DECLARATION OF STATE OF DISASTER

**23. Power of disaster district co-ordinator to declare state of disaster.**
(1) If at any time it appears to a disaster district co-ordinator that the magnitude or threatened magnitude of a disaster or impending disaster is or is likely to be so great in extent or severity in his disaster district that the counter-disaster measures necessary or desirable in respect of the disaster are or are likely to be beyond the capacity of the statutory services, he may, after consultation with the disaster district control group for that district and with the approval of the Minister, declare that a state of disaster exists in respect of that district.

(2) A declaration of a state of disaster made pursuant to this section –
(a) shall be in the prescribed form;
(b) shall come in force immediately after it is signed;
(c) shall continue in force for three days from and including the date the declaration was made unless sooner revoked by Order in Council.

(3) The duration of a state of disaster declared in a declaration made pursuant to this section may be extended by Order in Council for such periods not exceeding in each case 14 days as the Governor in Council thinks fit.

As amended by Act of 1978, No. 60, s. 7.

**24. Power of Governor in Council to declare state of disaster.**

(1) If at any time it is made to appear to the Governor in Council that the magnitude or threatened magnitude of a disaster or impending disaster is or is likely to be so great in extent or severity that the counter-disaster measures necessary or desirable with respect to the disaster are beyond the resources of a disaster district co-ordinator, he may, upon the recommendation of the Minister, by Order in Council, declare that a state of disaster exists in respect of any district or districts or of the whole State.

(2) (a) A declaration of a state of disaster made pursuant to this section-
(i) shall come into force immediately the Order in Council in respect thereof is made;
(ii) shall continue in force for 14 days from the date the Order in Council in respect was made unless sooner revoked.
(b) the duration of a state of disaster declared in a declaration pursuant to this section may be extended by Order in Council for such periods not exceeding in each case 14 days as the Governor in Council thinks fit.

(3) The Governor in Council by Order in Council –
(a) may at any time revoke a declaration made pursuant to this section or section 23;
(b) may extend for periods not exceeding in each case 14 days the duration of a state of disaster declared in a declaration pursuant to this section or section 23.

As amended by Act of 1978, no. 60, s. 8.

**25. Powers upon declaration of state of disaster.**

(1) Upon a declaration of a state of disaster pursuant to section 23 or 24 –

(a) the Chairman –
   (i) shall institute such measures as are in accordance with the disaster contingency plans, directions and orders of the Central Control Group;
   (ii) may direct the resources of the Government of the State and other resources that he considers necessary to relieve the effect of the disaster be made available;
   (iii) may take such other actions as he considers appropriate to counter the effects of the disaster;
   (iv) may authorize the expenditure of such sums of money as are determined by the Government of the State to relieve personal distress and assist in counter-disaster measures;
   (v) may if it appears to him that resources of the Commonwealth or of another State or Territory of the Commonwealth are necessary to make better provision for the relief of the effects of the disaster, with the approval of the Minister, request aid in accordance with the prescribed procedures.
(b) a disaster district co-ordinator –
   (i) shall take immediate action to use, direct and co-ordinate all available resources within his disaster district to counter the effects of the disaster;
   (ii) shall, immediately when they become available, use, direct and co-ordinate all additional resources that may be allocated or in respect of which arrangements have been made for allocation by the Central Control Group;
   (iii) shall take such other measures as are provided for in contingency plans and standing orders.

(2) Upon a declaration of a state of disaster pursuant to section 23 or 24 –
(a) the Chairman or a disaster district co-ordinator, if he is of opinion that such action is necessary for the preservation of human life and without any other authority other than this provision –
   (i) may require the owner or the person for the time being in charge of any resources to surrender them and place them under the control and direction of any authorized person involved in counter-disaster operations;
   (ii) may direct the evacuation and exclusion of persons from any place and in the exercise of those powers may remove or cause to be removed a person who does not comply with a direction to evacuate or a person who enters or is found in a place in respect of which a direction for the exclusion of persons has been given;

(b) a disaster district co-ordinator, local controller, police officer or an authorized person involved in counter-disaster operations without any authority other than this provision-

(i) may enter (by force if necessary) any place where he believes on reasonable grounds it is necessary so to do for the saving of human life or the prevention of injury to persons or for the rescue of injured or endangered persons or for facilitating the carrying out of other urgent measures with respect to the relief of suffering and distress;

(ii) may, if in his opinion it is necessary to do so for the conduct of counter-disaster operations, close to traffic any road, street, motorway, private street, private way, service lane, right of way or access way or other way or close any public place;

(iii) may remove from any place a vehicle or vessel that is impending counter-disaster operations and to facilitate its removal may use such force as is reasonably necessary or break into that vehicle or vessel.

(3) The powers conferred upon the Chairman or a disaster district coordinator by subsection (2) (a) (ii) may be exercised also by a local controller, police officer or an authorized person involved in counter-disaster operations.

(4) A person who suffers loss or damage to his property by reason of the exercise of the powers conferred by subsection (2) (a) (I) and subsection (2) (b) (iii) shall be entitled to claim in the manner prescribed compensation for such loss or damage.

A claim for compensation shall be made to the Minister whose decision thereon shall be final and conclusive.

As amended by Act of 1978, No. 60, s. 9.

## PART IV – POWERS AND DUTIES OF LOCAL AUTHORITIES

### 26. Duties as to counter-disaster measures.

(1) Save where the Local Authority has united with one or more than one other Local Authority for counter-disaster under subsection (2), each Local Authority –

(a) shall prepare a local counter-disaster plan to deal with all counter-disaster measures within its Area;

(b) shall establish and at all times maintain within its Area a local emergency service based upon the resources of the Local

Authority and including members of the State Emergency Service, volunteers and members of voluntary groups all of whom shall be resident in that Area, with such advisory committees, units and services as the Minister considers necessary or desirable;

(c) shall use, as prescribed or so far as not prescribed as the Minister determines, for counter-disaster purposes its resources and resources made available to it for those purposes.

(2) A Local Authority may, by agreement and with the approval of the Minister, unite with one or more than one other Local Authority for the purpose of arranging and carrying out counter-disaster measures in the combined Areas of the Local Authorities that are parties to the agreement.

(3) Where two or more Local Authorities have united for counter-disaster purposes under this section they –

(a) shall prepare a join counter-disaster plan to deal with all counter-disaster measures within the combined Areas;

(b) shall establish and at all times maintain within their combined Areas a local emergency service based upon the resources of the Combined Local Authorities and including members of the State Emergency Service, volunteers and members of voluntary groups all of whom shall be resident within those Areas, with such committees, units and services as the Minister considers necessary or desirable;

(c) shall use, as prescribed or so far as not prescribed as the Minister determines, for counter-disaster purposes the resources of the combined Local Authorities and resources made available to them for those purposes.

(4) Every counter-disaster plan prepared in accordance with this section shall be furnished by the Local Authority or, as the case requires, Combined Local Authorities to the appropriate disaster district coordinator for approval and shall upon approval form part of the disaster district plans of the Organization.

The procedure to be followed by a disaster district co-ordinator upon receipt by him of a plan furnished pursuant to this section shall be as prescribed.

(5) Upon the establishment of a local emergency service pursuant to this section, a Local Authority or, as the case requires, the Combined Local Authorities shall nominate a person to be the local controller and furnish that nomination to the Minister for approval and appointment, in accordance with the prescribed procedure.

A local controller shall have and may exercise such powers and perform such functions and duties as are prescribed or so far as not prescribed as the Minister determines.

As amended by Act of 1978, No. 60, s. 10.

27. **Power to unite in contracts for counter-disaster purposes.** Where two or more Local Authorities have united for counter-disaster purposes, they may for those purposes enter into contracts (being within their several powers) between themselves or some or one of them on behalf of all of them on the one hand generally may join in any act.

28. **Allocation to Local Authorities.**

(1) the Minister may, out of moneys appropriated by Parliament for the purposes of this Act, allocate to a Local Authority or Combined Local Authorities such sums and for such purposes as he determines.

(2) Sums allocated pursuant to this section shall be expended for the purposes for which the allocation was made.

## PART V – MISCELLANEOUS

29. **Protection from liability.** No action or proceeding shall lie or be brought or allowed by or in favor of any person against the Crown, the Minister, a Local Authority, Combined Local Authorities, members of the Police Force, the State Emergency Service of the Director or other officer thereof or any other person or body acting in the execution or intended execution of this Act or in accordance with any delegation under this Act or in compliance with any direction or purported to be given under this Act in respect of anything done or omitted to be done in good faith under and for the purposes of this Act.

30. **Offences with respect to officers and other persons.** A person shall not assault, obstruct, threaten, abuse, insult or intimidate the Director, an office or any other person in the exercise of his powers or the discharge of his functions and duties under this Act or attempt so to do.

31. **Offences generally and penalty.**

(1) A person who contravenes or fails to comply with any provision of this Act is guilty of an offence against this Act.

(2) A person who –

(a) fails to do that which is directed or required to do;

(b) does that which he is forbidden to do,

by a person acting under the authority of this Act commits an offense against this Act.

(3) A person who is guilty of an offence against this Act is liable to a penalty of $500.

**32. Proceeding of offences.** Prosecutions for offences against this Act shall be taken by way of summary proceedings under the Justices Act 1886-1975 upon the complaint of the Director or person authorized in writing in that behalf of the Minister.

**33. Offences by body corporate.**

(1) Where a body corporate commits an offence against this Act each of the following persons shall be deemed to have committed the offence and shall be liable to be proceeded against and punished accordingly: –

(a) the managing director, manager or other governing officer, by whatever name called and every member of the governing body thereof, by whatever name called; and

(b) every person who in Queensland manages or acts or takes part in the management, administration or government of the business in Queensland of the body corporate.

(2) This section applies so as not to limit or affect howsoever the liability of a body corporate to be proceeded against and punished for an offence against this Act committed by it.

**34. Evidentiary provisions.** In a proceeding for the purposes of this Act a certificate purporting to be signed by the Director and to certify –

(a) that a state of disaster was in existence either in the whole State or a part of the State as specified therein on a day or during a period specified therein;

(b) that a declaration of a state of disaster was in force on a day or during a period specified therein;

(c) that a delegation by the Minister, Chairman or Director to a person specified therein of the exercise or performance of any power, authority, function or duty specified therein was in force on a day or during a period specified therein;

(d) that a delegation by the Minister, Chairman or Director to a person specified therein was subject to the conditions or limitations specified therein or was not subject to any conditions or limitations at all; or

(e) that a person specified therein was a person authorized pursuant to this Act,

shall be evidence and, in the absence of evidence to the contrary, conclusive evidence of the facts so certified.

**34A. Annual Report.** The Director shall, as soon as practicable after the thirtieth day of June in each year, submit to the Minister a report on the administration and operations of the State of Emergency Service for the period of twelve months immediately prior to that date.

The Minister shall lay a copy of every such report before the Legislative Assembly.

Inserted by Act of 1978, No. 60, s. 11

**35. Protection of employment rights.** A person who during the period of a state of disaster declared pursuant to this Act is absent from usual employment on duties in connection with counter-disaster in any capacity whatever shall not be liable for dismissal, loss of long service leave, sick leave, recreation leave or other benefits to which he may be entitled under the industrial award applicable to his usual employment by reason only of his absence on those duties whether or not his usual employer has consented to his absence.

**36. Compensation for personal injury.** Every person who is a member of a local emergency service or any body acting under the authority of the Organization or the State Emergency Service shall, while he is engaged in counter-disaster operations or participating in counter-disaster training under the control of –

(a) a member of the Organization or a person acting under his authority;

(b) the Director or a person acting under his authority;

(c) a member of a local emergency service or a person acting under his authority,

be deemed to be a worker within the meaning of the *Worker's Compensation Act* 1916-1974 and the provision of that Act shall apply accordingly.

**37. Orders in Council.** Section 28 A of the *Acts Interpretation Act* 1954-1971 applies with respect to Orders in Council made for the purposes of this Act and, for the purposes of such application, that section shall be read and construed as if references to regulations therein were references to Orders in Council made for the purposes of this Act.

**38. Regulations.**

(1) The Governor in Council may make regulations not inconsistent with this Act for or with respect to all matters required or permitted

by this Act to be prescribed and all matters that in the opinion of the Governor in Council are necessary or convenient for the proper administration of this Act or to achieve the objects and purposes of this Act.

The regulations may prescribe penalties not exceeding in each case $100 for any contravention thereof or failure to comply therewith.

The power to regulate conferred by this subsection includes the power to prohibit.

(2) Regulations may be made –

(a) to apply generally or to meet a particular case or class of case;

(b) to apply throughout the State or within any part of the State.